BIG IDEAS MATH 7®

VIRGINIA EDITION

Record and Practice Journal

- Activity Recording Journal
- Activity Manipulatives
- Extra Practice Worksheets
- Fair Game Review Worksheets
- Glossary

BIG IDEAS LEARNING®

Erie, Pennsylvania

ISBN 13: 978-1-60840-167-3
ISBN 10: 1-60840-167-7

456789-VLP-14 13 12

Contents

Contents

Contents

Contents

Contents

Contents

Name____ Date__________

Chapter 1 Fair Game Review

Write a positive or negative integer that represents the situation.

1. You lose 5 points on a test.

2. A soccer team makes 8 goals.

3. You earn $47 raking leaves in your neighborhood.

4. The elevation of a submarine is 74 feet below sea level.

Use a number line to order the integers.

5. −9, 8, 0, 3, −7

6. −2, −4, 1, 2, −1

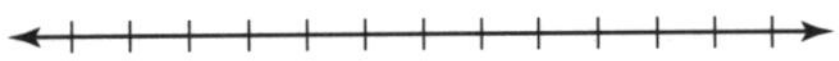

7. −11, −6, −8, 5, 9

8. 4, 2, −5, 0, −7

9. The table shows the lowest temperature recorded in January for five cities. Order the temperatures from least to greatest.

City	Boston	Nashville	Miami	Chicago	Dallas
Temperature (°F)	−12	−17	30	−27	4

Name ______________________________ Date ________

Chapter 1 Fair Game Review (continued)

Evaluate the expression.

10. $2 + 1 \bullet 4^2 - 12 \div 3$

11. $8^2 \div 16 \bullet 2 - 5$

12. $7(9 - 3) + 6^2 \bullet 10 - 8$

13. $3 \bullet 5 - 10 + 9(2 + 1)^2$

14. $8(6 + 5) - (9^2 + 3) \div 7$

15. $5[3(12 - 8)] - 6 \bullet 8 + 2^2$

16. $4 + 4 + 5 \times 2 \times 5 + (3 + 3 + 3) \times 6 \times 6 + 2 + 2$

a. Evaluate the expression.

b. Rewrite the expression using what you know about order of operations. Then evaluate.

Name______________________________ Date__________

1.1 Integers and Absolute Value

For use with Activity 1.1

Essential Question How are velocity and speed related?

On these two pages, you will investigate vertical motion (up or down).

- Speed tells how fast an object is moving, but does not tell the direction.
- Velocity tells how fast an object is moving and also tells the direction.

  If velocity is positive, the object is moving up.

  If velocity if negative, the object is moving down.

1 EXAMPLE: Falling Parachute

You are gliding to the ground wearing a parachute. The table shows your height at different times.

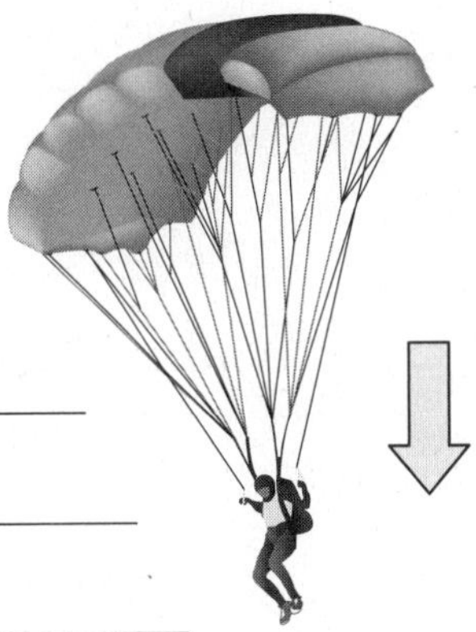

Time (seconds)	0	1	2	3
Height (feet)	45	30	15	0

a. How many feet do you move each second? ______________

b. What is your speed? Give the units. ______________

c. Is your velocity *positive* or *negative*? ______________

d. What is your velocity? Give the units. ______________

2 ACTIVITY: Rising Balloons

Work with a partner. The table shows the height of a group of balloons.

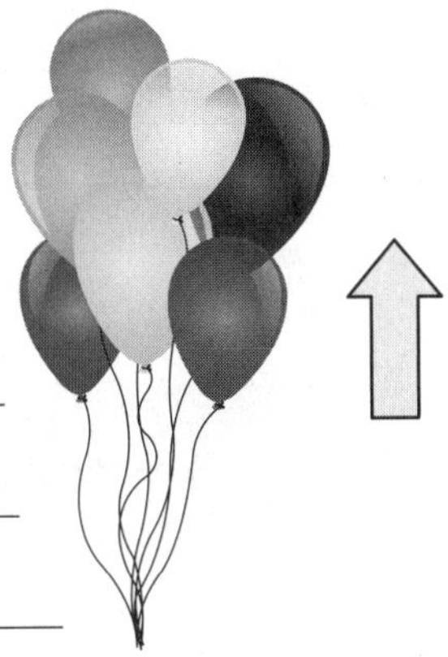

Time (seconds)	0	1	2	3
Height (feet)	0	4	8	12

a. How many feet do the balloons move each second? ______________

b. What is the speed of the balloons? Give the units. ______________

c. Is the velocity *positive* or *negative*? ______________

d. What is the velocity? Give the units. ______________

Name ______________________________ Date __________

1.1 Integers and Absolute Value (continued)

3 ACTIVITY: Finding Speed and Velocity

Work with a partner. The table shows the height of a firework's parachute.

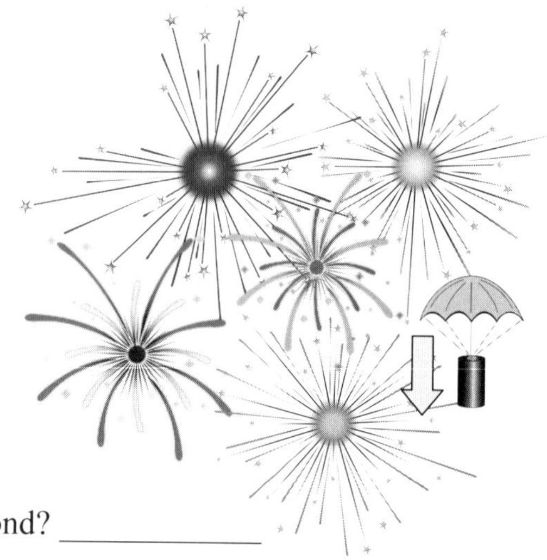

Time (seconds)	Height (feet)
0	480
1	360
2	240
3	120
4	0

a. How many feet does the parachute move each second? ____________

b. What is the speed of the parachute? Give the units. ____________

c. Is the velocity *positive* or *negative*? ____________

d. What is the velocity? Give the units. ____________

Inductive Reasoning

4. Complete the table.

Velocity (feet per second)	–14	20	–2	0	25	–15
Speed (feet per second)						

5. Find two different velocities for which the speed is 16 feet per second.

6. Which number is greater: –4 or 3?

Use a number line to explain your reasoning.

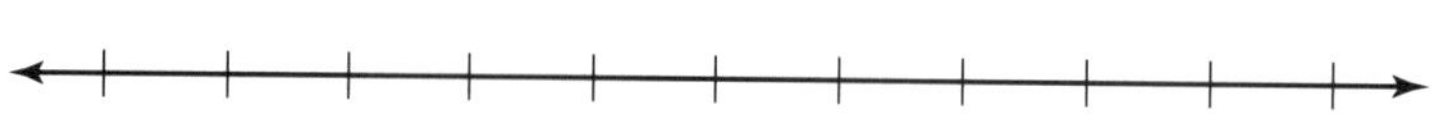

7. One object has a velocity of -4 feet per second. Another object has a velocity of 3 feet per second. Which object has the greater speed? Explain your answer.

What Is Your Answer?

In this lesson, you will study **absolute value**. Here are some examples:

Absolute value of $-16 = 16$ Absolute value of $16 = 16$

Absolute value of $0 = 0$ Absolute value of $-2 = 2$

8. IN YOUR OWN WORDS How are velocity and speed related?

9. Which of the following is a true statement? Explain your reasoning.

a. Absolute value of velocity = speed

b. Absolute value of speed = velocity

Name ____________________ Date ________

1.1 Practice
For use after Lesson 1.1

Find the absolute value.

1. $|-1|$ **2.** $|-14|$ **3.** $|0|$ **4.** $|6|$

Complete the statement using <, >, or =.

5. 6 ___ $|-2|$ **6.** -7 ___ $|-8|$ **7.** $|-9|$ ___ 5 **8.** $|-2|$ ___ 2

Order the values from least to greatest.

9. $4,\ |7|,\ -1,\ |-3|,\ -4$ **10.** $|2|,\ -3,\ |-5|,\ -1,\ 6$ **11.** $|-8|,\ 0,\ -9,\ |-7|,\ -2$

12. You download 12 new songs to your MP3 player. Then you delete 5 old songs. Write each amount as an integer.

13. The mantle layer of Earth begins about 70 kilometers underneath the surface. The layer called the outer core begins about 2970 kilometers underneath the surface.

a. Write an integer for the position of each layer relative to the surface.

Mantle layer ____________ Outer core ____________

b. Which integer in part (a) is greater?

c. Which integer in part (a) has the greater absolute value? How does this relate to the layer that has the greatest distance from the surface?

Name____ Date__________

1.2 Adding Integers

For use with Activity 1.2

Essential Question Is the sum of two integers *positive*, *negative*, or *zero*? How can you tell?

1 EXAMPLE: Adding Integers with the Same Sign

Draw a picture to show how you use integer counters to find $-4 + (-3)$.

$-4 + (-3) =$ ______

2 ACTIVITY: Adding Integers with Different Signs

Work with a partner. Draw a picture to show how you use integer counters to find $-3 + 2$.

$-3 + 2 =$ ______

3 EXAMPLE: Adding Integers with Different Signs

Show how to use a number line to find $5 + (-3)$.

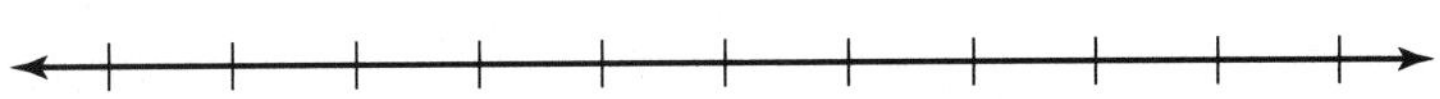

$5 + (-3) =$ ______

1.2 Adding Integers (continued)

4 ACTIVITY: Adding Integers with Different Signs

Work with a partner. Write the addition expression shown. Then find the sum.

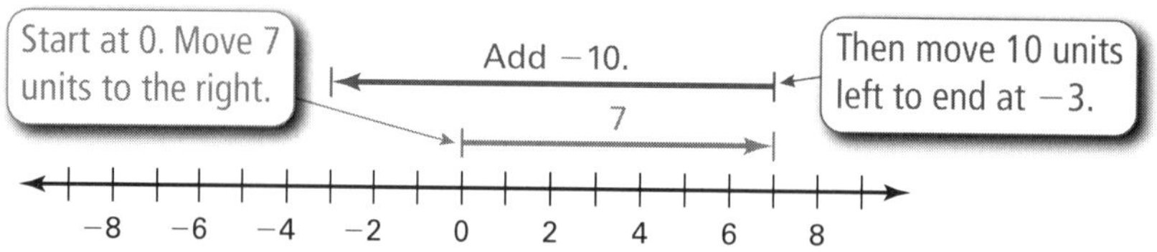

Inductive Reasoning

Work with a partner. Use integer counters or a number line to complete the table.

	Exercise	Type of Sum	Sum	Sum: Positive, Negative, or Zero
1	**5.** $-4 + (-3)$			
2	**6.** $-3 + 2$			
3	**7.** $5 + (-3)$			
4	**8.** $7 + (-10)$			
	9. $2 + 4$			
	10. $-6 + (-2)$			
	11. $-5 + 9$			
	12. $15 + (-9)$			
	13. $-10 + 10$			
	14. $-6 + (-6)$			
	15. $12 + (-12)$			

Name____ Date__________

What Is Your Answer?

16. **IN YOUR OWN WORDS** Is the sum of two integers *positive*, *negative*, or *zero*? How can you tell?

17. Write a general rule for adding

    a. two integers with the same sign.

    b. two integers with different signs.

    c. an integer and its opposite.

Name ______________________________ Date __________

1.2 Practice

For use after Lesson 1.2

Add.

1. $-9 + 2$

2. $5 + (-5)$

3. $-12 + (-6)$

4. $-10 + 19 + 5$

5. $-11 + (-20) + 9$

6. $-7 + 7 + (-8)$

Use mental math to solve the equation.

7. $x + (-5) = 4$

8. $y + 6 = -2$

9. $-10 = -7 + z$

10. You have –30 points in a video game. You earn 57 points. What is your final score?

11. The table shows the change in your hair length over a year.

Month	January	February	August	September	December
Change in hair length (inches)	2	–1	3	–4	3

a. What is the total change in your hair length at the end of the year?

b. Is your hair longer in January or December? Explain your reasoning.

c. When is your hair the longest? Explain your reasoning.

Name____ Date __________

1.3 Subtracting Integers

For use with Activity 1.3

Essential Question How are adding integers and subtracting integers related?

1 EXAMPLE: Subtracting Integers

Draw a picture to show how you use integer counters to find $4 - 2$.

$4 - 2 =$ _______

2 ACTIVITY: Adding Integers

Work with a partner. Draw a picture to show how you use integer counters to find $4 + (-2)$.

$4 + (-2) =$ _______

3 EXAMPLE: Subtracting Integers

Show how to use a number line to find $-3 - 1$.

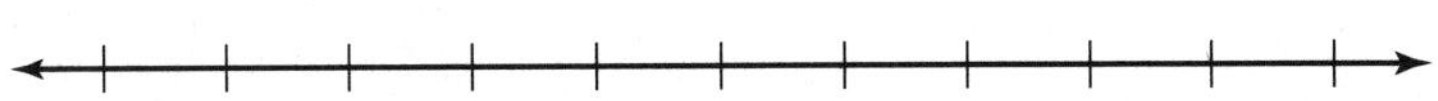

$-3 - 1 =$ _______

Name ______________________________ Date __________

1.3 Subtracting Integers (continued)

4 ACTIVITY: Adding Integers

Work with a partner. Write the addition expression shown. Then find the sum.

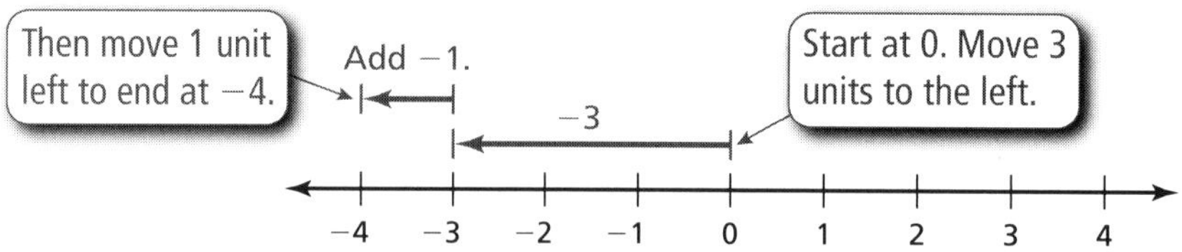

Inductive Reasoning

Work with a partner. Use integer counters or a number line to complete the table.

	Exercise	Operation: Add or Subtract	Answer
1	**5.** $4 - 2$		
2	**6.** $4 + (-2)$		
3	**7.** $-3 - 1$		
4	**8.** $-3 + (-1)$		
	9. $3 - 8$		
	10. $3 + (-8)$		
	11. $9 - 13$		
	12. $9 + (-13)$		
	13. $-6 - (-3)$		
	14. $-6 + 3$		
	15. $-5 - (-12)$		
	16. $-5 + 12$		

1.3 Subtracting Integers (continued)

What Is Your Answer?

17. **IN YOUR OWN WORDS** How are adding integers and subtracting integers related?

18. Write a general rule for subtracting integers.

Name ______________________________ Date __________

1.3 Practice
For use after Lesson 1.3

Subtract.

1. $3 - 8$ **2.** $6 - (-7)$ **3.** $-10 - 9$ **4.** $-5 - (-4)$

Evaluate the expression.

5. $11 - (-2) + 14$ **6.** $-16 - (-12) + (-8)$ **7.** $6 - 17 - 4$

Use mental math to solve the equation.

8. $6 - x = 10$ **9.** $y - (-10) = 2$ **10.** $z - 17 = -14$

11. You begin a hike in Death Valley, California at an elevation of –86 meters. You hike to a point of elevation at 45 meters. What is your change in elevation?

12. You sell T-shirts for a fundraiser. It costs \$112 to have the T-shirts made. You make \$98 in sales. What is your profit?

13. The table shows the scores of six golfers. Find the range of the scores.

Golfer	1	2	3	4	5	6
Score	–5	–2	3	–3	1	–8

Name____ Date__________

1.4 Multiplying Integers

For use with Activity 1.4

Essential Question Is the product of two integers *positive*, *negative*, or *zero*? How can you tell?

1 EXAMPLE: Multiplying Integers with the Same Sign

Use repeated addition to find $3 \bullet 2$.

Recall that multiplication is repeated addition. $3 \bullet 2$ means to add 3 groups of 2.

$3 \bullet 2 =$ _______

2 EXAMPLE: Multiplying Integers with Different Signs

Use repeated addition to find $3 \bullet (-2)$.

$3 \bullet (-2) =$ _______

3 ACTIVITY: Multiplying Integers with Different Signs

Work with a partner. Use a table to find $-3 \bullet 2$.

Describe the pattern in the table. Use the pattern to complete the table.

$2 \bullet 2 =$
$1 \bullet 2 =$
$0 \bullet 2 =$
$-1 \bullet 2 =$
$-2 \bullet 2 =$
$-3 \bullet 2 =$

$-3 \bullet 2 =$ _______

Name ______________________________ Date __________

4 ACTIVITY: Multiplying Integers with the Same Sign

Work with a partner. Use a table to find $-3 \bullet (-2)$.

Describe the pattern in the table. Use the pattern to complete the table.

−3	•	3	=
−3	•	2	=
−3	•	1	=
−3	•	0	=
−3	•	−1	=
−3	•	−2	=

$-3 \bullet (-2) =$ ______

Inductive Reasoning

Work with a partner. Complete the table.

	Exercise	Type of Product	Product	Product: Positive or Negative
1	**5.** $3 \bullet 2$			
2	**6.** $3 \bullet (-2)$			
3	**7.** $-3 \bullet 2$			
4	**8.** $-3 \bullet (-2)$			
	9. $6 \bullet 3$			
	10. $2 \bullet (-5)$			
	11. $-6 \bullet 5$			
	12. $-5 \bullet (-3)$			

13. Write two integers whose product is 0.

What Is Your Answer?

14. **IN YOUR OWN WORDS** Is the product of two integers *positive*, *negative*, or *zero*? How can you tell?

15. Write a general rule for multiplying

   a. two integers with the same sign.

   b. two integers with different signs.

Name ______________________________ Date ________

1.4 Practice
For use after Lesson 1.4

Multiply.

1. $8 \bullet 9$ **2.** $7(-7)$ **3.** $-10 \bullet 4$ **4.** $-5(-6)$

5. $12 \bullet (-1) \bullet (-2)$ **6.** $-10(-3)(-7)$ **7.** $-20 \bullet 0 \bullet (-4)$ **8.** $-4 \bullet 8 \bullet 3$

Evaluate the expression.

9. $(-8)^2$ **10.** -11^2 **11.** $9 \bullet (-5)^2$ **12.** $(-2)^3 \bullet (-6)$

13. You lose 5 points for every wrong answer in a trivia game. What integer represents the change in your points after answering 8 questions wrong?

14. A glacier is melting at a rate of 36 cubic miles each year. What integer represents the change in the glacier's volume over 4 years?

15. The value of a computer is given by the expression $2000 + (-200t)$, where t is the time in years.

a. Complete the table.

Time	1 year	2 years	3 years	4 years
Value				

b. Describe the change in the value of the computer for each year.

1.5 Dividing Integers

For use with Activity 1.5

Essential Question Is the quotient of two integers *positive*, *negative*, or *zero*? How can you tell?

1 EXAMPLE: Dividing Integers with Different Signs

Draw a picture to show how you use integer counters to find $-15 \div 3$.

$-15 \div 3 =$ _______

2 ACTIVITY: Rewriting a Product as a Quotient

Work with a partner. Rewrite the product $3 \bullet 4 = 12$ as a quotient in two different ways.

First Way

12 is equal to 3 groups of _______.

$12 \div 3 =$ _______

Second Way

12 is equal to 4 groups of _______.

$12 \div 4 =$ _______

3 EXAMPLE: Dividing Integers with Different Signs

Rewrite the product $-3 \bullet (-4) = 12$ as a quotient in two different ways. What can you conclude?

First Way

Second Way

Name ______________________________ Date __________

1.5 Dividing Integers (continued)

4 EXAMPLE: Dividing Negative Integers

Rewrite the product $3 \bullet (-4) = -12$ as a quotient in two different ways. What can you conclude?

First way *Second Way*

Inductive Reasoning

Work with a partner. Complete the table.

	Exercise	Type of Quotient	Quotient	Quotient: Positive, Negative, or Zero
1	**5.** $-15 \div 3$			
2	**6.** $12 \div 4$			
3	**7.** $12 \div (-3)$			
4	**8.** $-12 \div (-4)$			
	9. $-6 \div 2$			
	10. $-21 \div (-7)$			
	11. $10 \div (-2)$			
	12. $12 \div (-6)$			
	13. $0 \div (-15)$			
	14. $0 \div 4$			

What Is Your Answer?

15. IN YOUR OWN WORDS Is the quotient of two integers *positive*, *negative*, or *zero*? How can you tell?

16. Write a general rule for dividing

a. two integers with the same sign.

b. two integers with different signs.

Name ______________________________ Date __________

1.5 Practice

For use after Lesson 1.5

Divide, if possible.

1. $3 \div (-1)$

2. $8 \div 2$

3. $-10 \div 5$

4. $-21 \div (-7)$

5. $\frac{48}{-6}$

6. $\frac{-13}{-13}$

7. $\frac{0}{3}$

8. $\frac{-55}{11}$

Evaluate the expression.

9. $-63 \div (-7) + 6$

10. $-5 - 12 \div 3$

11. $-8 \bullet 7 + 33 \div (-11)$

12. An online group loses 20 members over five months. What is the mean monthly change in the group membership?

13. The table shows the number of yards a football player runs in each quarter of a game. Find the mean number of yards the player runs per quarter.

Quarter	1	2	3	4
Yards	–2	14	–18	–6

14. The far north region of Alaska has an average temperature of –22°F during the month of March. The interior region has an average temperature of –2°F during the month of March. How many times colder is the far north region than the interior region during March?

Name______________________________ Date__________

Chapter 2 Fair Game Review

Find the area of the polygon.

1.

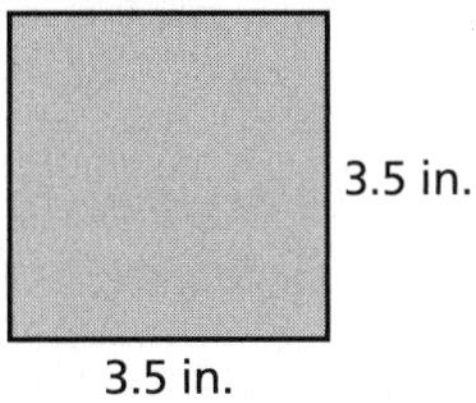

2.

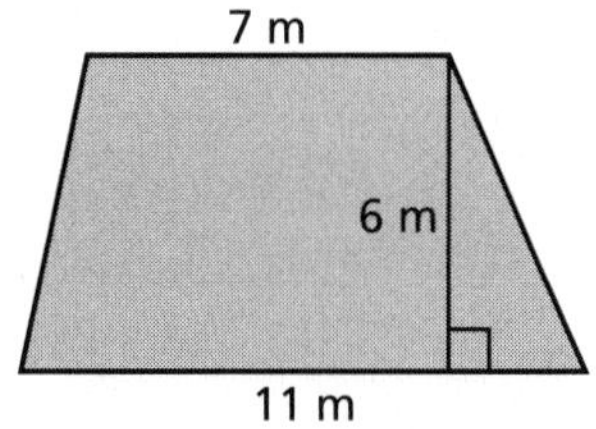

3.

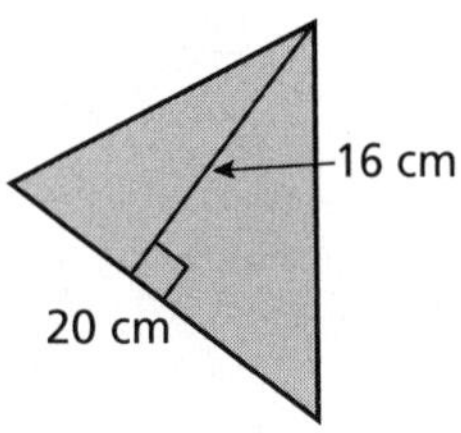

4.

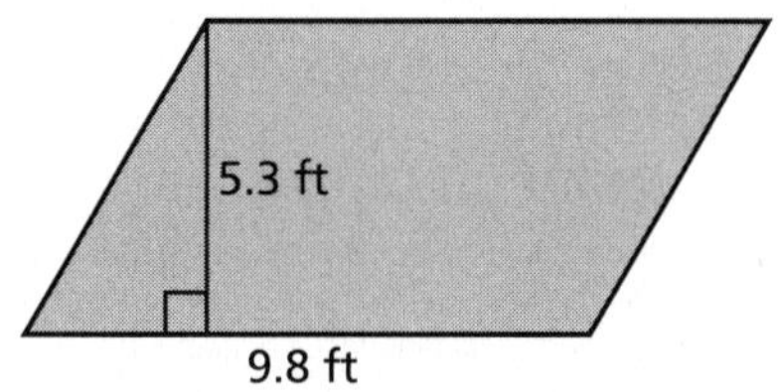

5.

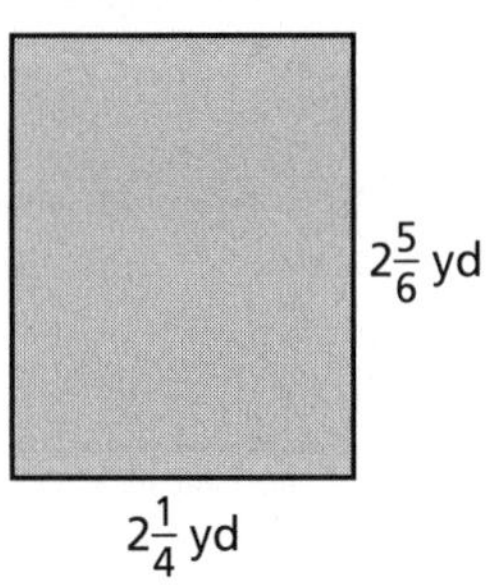

6.

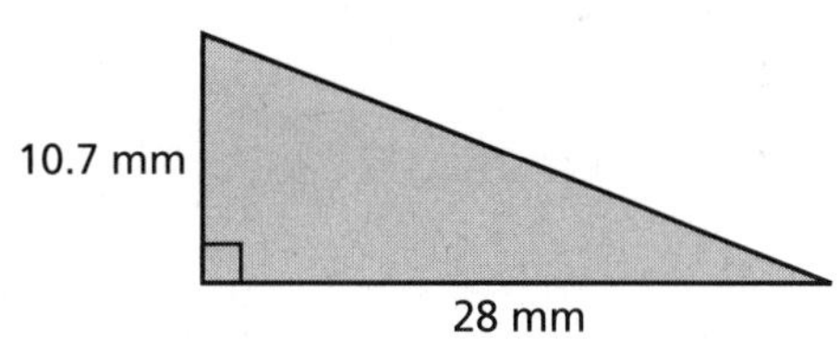

7. You are painting a trapezoidal portion of a backdrop for the school play. The base lengths of the trapezoid are 3 feet and 6 feet. The height is 3 feet. What is the area of the portion you are painting?

Name ______________________________ Date ________

Chapter 2 Fair Game Review (continued)

Use the Distributive Property to simplify the expression.

8. $4(x + 7)$

9. $\frac{1}{8}(24 - g)$

10. $3(2.1 + n)$

11. $5(w - 1.9)$

12. $9(4 + z + 6)$

13. $10(0.3 - b - 7.4)$

14. $\frac{1}{6}(30 + m - 18)$

15. $7.5(0.8 - y + 3.2)$

16. You and four friends go to the movies. Each of you pays \$3 for popcorn, \$$x$ for a ticket, and \$2 for a drink. Simplify the expression $5(3 + x + 2)$ that represents your total cost at the movies.

Name ______________________________ Date __________

2.1 Evaluating Algebraic Expressions

For use with Activity 2.1

Essential Question What does it mean to *evaluate* an expression that has variables?

1 ACTIVITY: Rate and Time Problems

Work with a partner. Many real-life uses of mathematics involve equations of the form

Amount = Rate • Time.

Write and solve a real-life question for each "rate-time" problem. Illustrate your question with a drawing, diagram, or graph.

a. $D = r \bullet t$

Distance	=	Rate	•	Time
km		km per h		h

b. $D = r \bullet t$

Distance	=	Rate	•	Time
ft		ft per sec		sec

c. $P = w \bullet t$

Pay	=	Wage	•	Time
$		$ per h		h

Name ______________________________ Date __________

2.1 Evaluating Algebraic Expressions (continued)

d. $P = r \bullet t$

Population	=	Rate	•	Time
people		people per yr		yr

e. $C = r \bullet t$

Rental Cost	=	Rate	•	Time
$		$ per day		day

2 ACTIVITY: Solving Puzzles

Work with a partner. Complete each puzzle using each number from 1 through 9 *exactly once*. Read each row from left to right and each column from top to bottom.

Note: An arrow from one dashed circle to another indicates that the same number goes in both circles.

a.

5	×	2	÷		= 10
−	■	+	■	×	
	÷		+		= 11
+	■	×	■	−	
◌	×		−	◌	= 17
= 0		= 20		= 1	

b.

	÷		+		= 13
−	■	×	■	+	
	×		×		= 30
+	■	÷	■	−	
◌	×		+	◌	= 25
= 14		= 4		= 10	

c.

◌	−		÷	◌	= 1
×	■	+	■	+	
	×		+		= 11
+	■	−	■	×	
	+		×		= 68
= 9		= 5		= 23	

d.

◌	+		÷	◌	= 4
×	■	×	■	+	
	+		÷		= 8
÷	■	+	■	×	
	×		−		= 1
= 10		= 79		= 22	

Name______________________________ Date__________

e. Make up your own puzzle. (Write the answers on a separate piece of paper.) Trade puzzles with your partner and try solving each other's puzzle.

What Is Your Answer?

3. **IN YOUR OWN WORDS** What does it mean to *evaluate* an expression that has variables? Give an example to illustrate your answer.

Name ______________________________ Date __________

2.1 Practice

For use after Lesson 2.1

Evaluate the expression when $a = -5$, $b = 2.4$, and $c = 36$.

1. $\frac{c}{6}$

2. $b - 3.1$

3. $12a$

4. $b - a$

5. $\frac{c}{b}$

6. $c + a$

7. ab

8. $\frac{c}{ab}$

9. $3b - 7.2$

10. $a^2 + 15$

11. $-7 + c \div b$

12. $\frac{b^2 + 4a}{4}$

13. You are $a - 5$ years old. Your sister is a years old. How old are you when your sister is 21 years old?

14. The expression $4.5p + 12h + b$ represents the cost of buying p paperback books, h hardback books, and b bookmarks. How much do you spend when you buy 3 paperback books, 2 hardback books, and 5 bookmarks?

Name____ Date__________

2.2 Writing Expressions and Equations

For use with Activity 2.2

Essential Question How can you use math, logic, and diagrams to solve word problems?

1 ACTIVITY: Solving Brain Teasers

Work with a partner. When you solve real-life problems that involve math, communication and clear writing are important. Here are some brain teasers and puzzles that illustrate this. Solve each brain teaser.

a. You have two U.S. coins whose value is $0.55. One of them is not a nickel. What are the coins?

b. One brick weighs one pound plus half a brick. How much, in pounds, does one brick weigh?

c. You drive 60 miles to town at a speed of 30 miles per hour. How fast do you have to drive on the return trip so that your average round-trip speed is 60 miles per hour?

d. If it were two hours later, it would be half as long until midnight as it would be if it were an hour later. What time is it now?

Name ____ Date __________

2.2 Writing Expressions and Equations (continued)

e. You and your friend live in different sections of town but go to the same school. You left for school ten minutes before your friend and you meet in the park. When you meet, who is closer to school?

2 ACTIVITY: Writing a Brain Teaser

Work with a partner.

Write your own brain teaser. Make it fun and interesting. Use color and a drawing to make it appealing.

When you are finished, trade brain teasers with your classmates and see who can solve the brain teasers quickest.

Name____ Date__________

2.2 Writing Expressions and Equations (continued)

3 ACTIVITY: Brain Teasers in History

In about 1650 B.C., the Egyptian scribe Ahmes made a transcript of an even more ancient mathematical manuscript dating to the reign of the Pharaoh Amenemhat III. The manuscript is filled with math riddles. Here is one that is similar to a riddle that was in the manuscript. Work with your partner to solve this riddle.

One hundred measures of corn must be divided among 5 workers. The third, fourth, and fifth workers together get four times as much as the first and second workers together. The second worker gets 4 times as much as the first worker. The third and fourth workers get equal amounts. The fifth worker gets twice as much as the third worker. How many measures of corn does each worker get?

What Is Your Answer?

4. **IN YOUR OWN WORDS** How can you use math, logic, and diagrams to solve word problems?

Name ______________________________ Date ________

2.2 Practice

For use after Lesson 2.2

Write the phrase as an expression.

1. a number c subtracted from 18

2. 7 plus the product of 4 and a number x

Write the word sentence as an equation.

3. A number k increased by 13 equals 5.

4. The quotient of a number b and 3.2 is 20.

5. Negative 9 is 8.5 subtracted from one-half of a number y.

6. 30 more than a number g divided by 2 equals 24.

Write an equation that can be used to find the value of *x*.

7. Area of a parallelogram: 36 in.^{2}

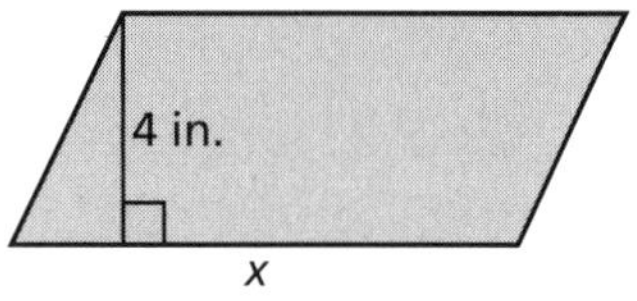

8. Circumference of a circle: 17π m

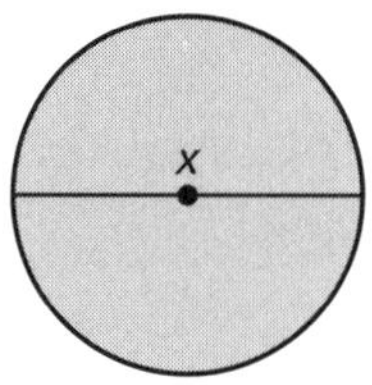

9. You have a coupon for 15% off any item in the store. Write an expression for the amount you save using the coupon.

10. A triangle is added to a rectangle. The area of the figure is 75 square inches. Write an equation that can be used to find the missing measurement x.

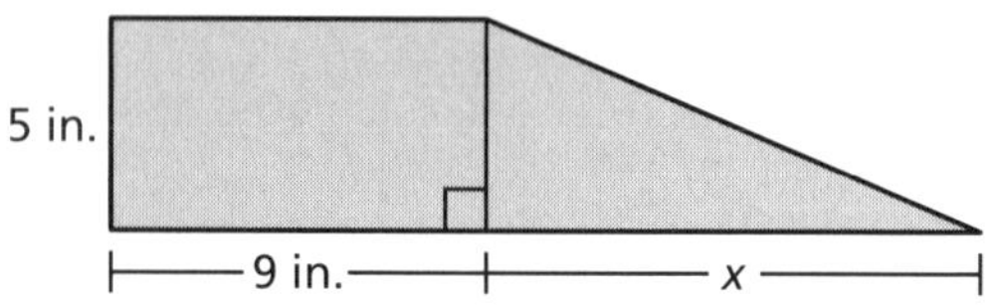

Name____ Date__________

2.3 Using Formulas to Solve Problems

For use with Activity 2.3

Essential Question How can you use formulas to find the area of an object with an unusual shape?

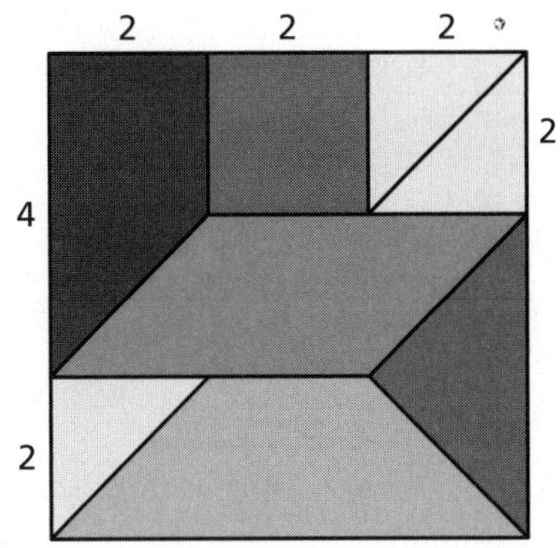

1 ACTIVITY: Using an Area Formula

Work with a partner. Complete the table.

Polygon	Name	Area Formula	Area
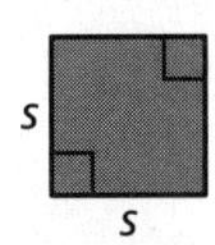			
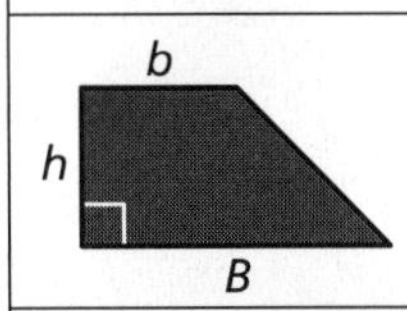			
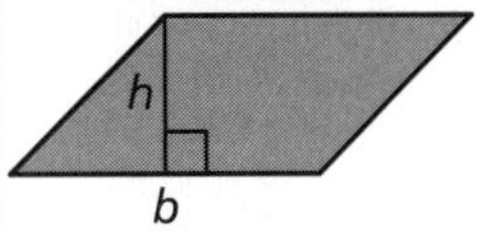			
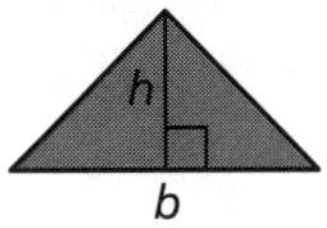			
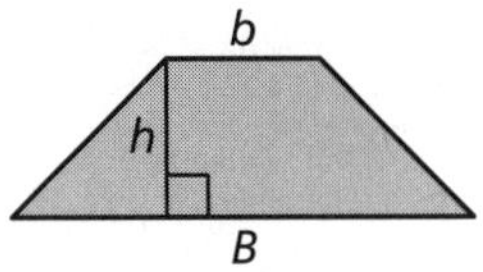			
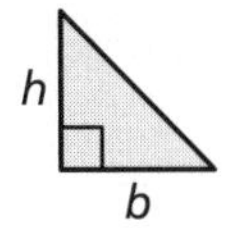			

Name ______________________________ Date __________

2 ACTIVITY: Finding an Area

Work with a partner. Use the shapes from Activity 1 to find the area of the sailboat. Explain your reasoning.

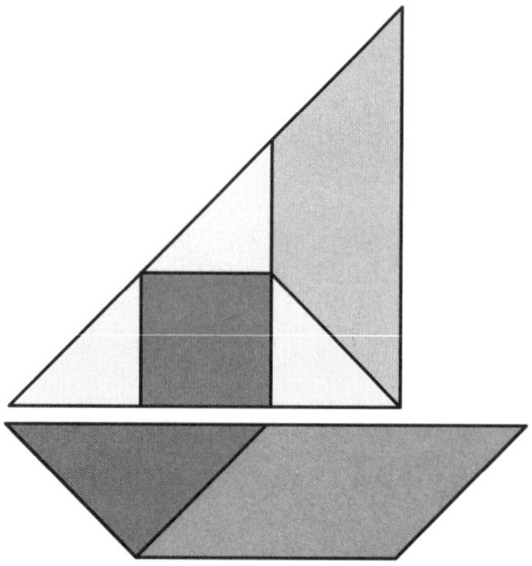

3 ACTIVITY: Finding an Area

Work with a partner. Use the shapes from Activity 1 to create the picture. Then draw the outline of each shape on the figure to show how you created the picture.*

a. house

20 square units

b. rabbit

36 square units

c. bird

32 square units

*Cut-outs are available in the back of the Record and Practice Journal.

2.3 Using Formulas to Solve Problems (continued)

What Is Your Answer?

4. IN YOUR OWN WORDS How can you use formulas to find the area of an object with an unusual shape?

5. Show how you can use the formula $A = bh$ for the area of a rectangle to write the formula for the area of a parallelogram.

6. Show how you can use the formula $A = bh$ for the area of a rectangle to write the formula for the area of a triangle.

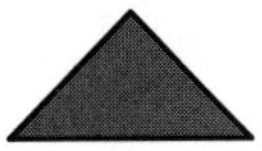

Name ____________________ Date __________

2.3 Practice
For use after Lesson 2.3

Use a formula to find the area of the figure.

1. 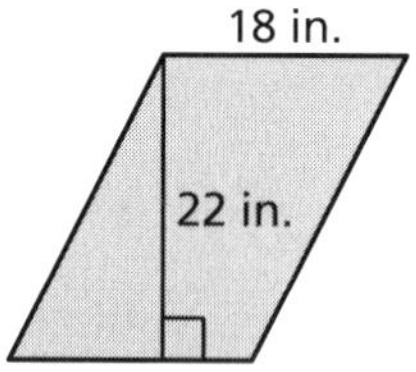

2. 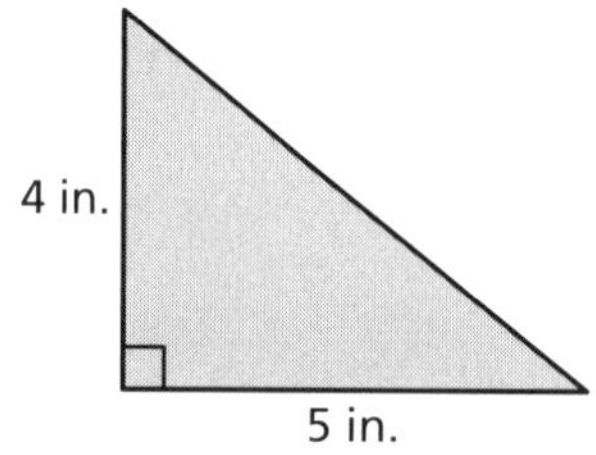

3. 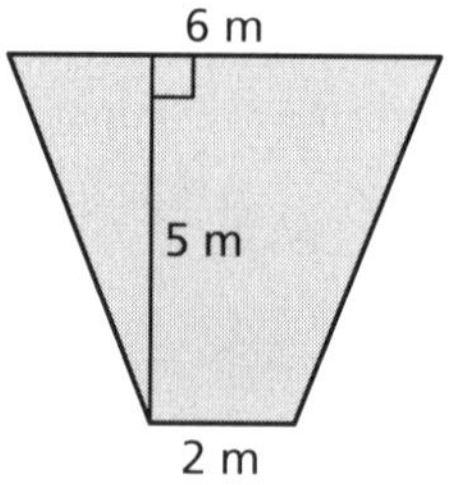

4. In football, a player's average number of yards per carry is the average number of yards gained divided by the number of times the player carries the ball. What is the average number of yards per carry for a player who gained 156 yards on 12 carries?

$$\text{Average number of yards per carry} = \frac{\text{Number of yards gained}}{\text{Number of carries}}$$

5. A rectangular lawn is 60 feet wide and 90 feet long.

a. What is the area of the lawn?

b. A bag of fertilizer covers 50 square feet. How many bags of fertilizer do you need to cover the entire lawn? Explain.

Name ______________________________ Date __________

2.4 Simplifying Algebraic Expressions

For use with Activity 2.4

Essential Question How can you simplify an algebraic expression?

1 ACTIVITY: Simplifying Algebraic Expressions

Work with a partner. Evaluate each expression when $x = 0$ and when $x = 1$. Use the results to match each expression in the left column with its equivalent expression in the right column.

	Expression	Value When $x = 0$	Value When $x = 1$
A.	$3x + 2 - x + 4$		
B.	$5(x - 3) + 2$		
C.	$x + 3 - (2x + 1)$		
D.	$-4x + 2 - x + 3x$		
E.	$-(1 - x) + 3$		
F.	$2x + x - 3x + 4$		
G.	$4 - 3 + 2(x - 1)$		
H.	$2(1 - x + 4)$		
I.	$5 - (4 - x + 2x)$		
J.	$5x - (2x + 4 - x)$		

	Expression	Value When $x = 0$	Value When $x = 1$
a.	4		
b.	$-x + 1$		
c.	$4x - 4$		
d.	$2x + 6$		
e.	$5x - 13$		
f.	$-2x + 10$		
g.	$x + 2$		
h.	$2x - 1$		
i.	$-2x + 2$		
j.	$-x + 2$		

2.4 Simplifying Algebraic Expressions (continued)

2 ACTIVITY: Writing a Math Lesson

Work with a partner. Use your results from Activity 1 to write a lesson on simplifying an expression.

Simplifying Expressions

Key Idea Use the following steps to simplify an algebraic expression.

1.

2.

3.

Describe steps you can use to simplify an expression.

Example 1

Write 3 examples. Use expressions from Activity 1.

a.

b.

c.

Exercises

Write 3 exercises. Use expressions different from the ones in Activity 1.

Simplify the expression.

1.

2.

3.

What Is Your Answer?

3. **IN YOUR OWN WORDS** How can you simplify an algebraic expression? Give an example that demonstrates your procedure.

4. **RESEARCH** Use the Internet or some other source to find an example of an algebraic expression in real life. Is the expression simplified? Explain why or why not.

Name ______________________ Date __________

2.4 Practice

For use after Lesson 2.4

Identify the terms and like terms in the expression.

1. $3x + 4 - 7x - 6$

2. $-9 + 2.5y - 0.7y + 1 + 6.4y^2$

Simplify the expression.

3. $5a - 2a + 9$

4. $m - \frac{1}{6} - 4m + \frac{5}{6}$

5. $2.3w - 7 + 8.1 - 3w$

6. $7(d - 1) + 2$

7. $13.7g + 2(4k - 1.5g)$

8. $20(p + 2) + 16(-3 - p)$

9. $\frac{1}{2}u^2 + \frac{1}{4}u^2 - \frac{1}{8}t$

10. $3(4s - s^2) + s^2 - 5$

11. $12z^2 - 9n + 15n - 31z^2$

12. Write an expression in simplest form that represents the perimeter of the polygon.

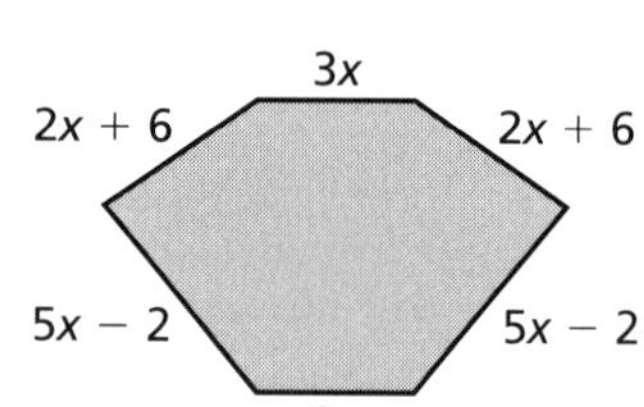

13. Write an expression in simplest form that represents the cost for shampooing and cutting *w* women's hair and *m* men's hair.

	Women	**Men**
Cut	\$15	\$7
Shampoo	\$5	\$2

Name____ Date__________

Chapter 3 Fair Game Review

Write the decimal as a fraction.

1. 0.26

2. 0.79

3. 0.571

4. 0.846

Write the fraction as a decimal.

5. $\frac{3}{8}$

6. $\frac{4}{10}$

7. $\frac{11}{16}$

8. $\frac{17}{20}$

9. A quarterback completed 0.6 of his passes during a game. Write the decimal as a fraction.

Name ______________________________ Date __________

Chapter 3 Fair Game Review (continued)

Evaluate the expression.

10. $\frac{1}{8} + \frac{1}{9}$

11. $\frac{2}{3} + \frac{9}{10}$

12. $\frac{7}{12} - \frac{1}{4}$

13. $\frac{6}{7} - \frac{4}{5}$

14. $\frac{5}{9} \bullet \frac{1}{3}$

15. $\frac{8}{15} \bullet \frac{3}{4}$

16. $\frac{7}{8} \div \frac{11}{16}$

17. $\frac{3}{10} \div \frac{2}{5}$

18. You have 8 cups of flour. A recipe calls for $\frac{2}{3}$ cup of flour. Another recipe calls for $\frac{1}{4}$ cup of flour. How much flour do you have left after making the recipes?

Name____ Date__________

3.1 Rational Numbers

For use with Activity 3.1

Essential Question How can you use a number line to order rational numbers?

A **rational number** is a number that can be written as a ratio of two integers.

$$2 = \frac{2}{1} \qquad -3 = \frac{-3}{1} \qquad -\frac{1}{2} = \frac{-1}{2} \qquad 0.25 = \frac{1}{4}$$

1 ACTIVITY: Ordering Rational Numbers

Work in groups of five. Order the numbers from least to greatest.

a. $-0.5, 1.25, -\frac{1}{3}, 0.5, -\frac{5}{3}$

- Make a number line on the floor using masking tape and a marker.

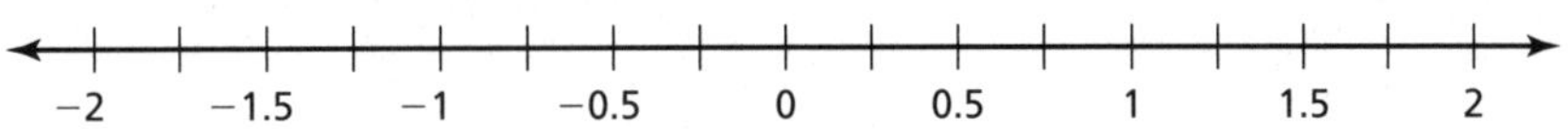

- Write the numbers on pieces of paper. Then each person should choose one piece of paper.
- Stand on the location of your number on the number line.
- Use your positions to order the numbers from least to greatest.

  The numbers from least to greatest are

  ______, ______, ______, ______, and ______.

b. $-\frac{7}{4}, 1.1, \frac{1}{2}, -\frac{1}{10}, -1.3$

c. $-\frac{1}{4}, 2.5, \frac{3}{4}, -1.7, -0.3$

d. $-1.4, -\frac{3}{5}, \frac{9}{2}, \frac{1}{4}, 0.9$

e. $\frac{9}{4}, 0.75, -\frac{5}{4}, -0.8, -1.1$

Name ______________________________ Date __________

2 ACTIVITY: The Game of Math Card War

Preparation:

- Cut index cards to make 40 playing cards.
- Write each number in the table on a card.

$-\frac{3}{2}$	$\frac{3}{10}$	$-\frac{3}{4}$	−0.6	1.25	−0.15	$\frac{5}{4}$	$\frac{3}{5}$	−1.6	−0.3
$\frac{3}{20}$	$\frac{8}{5}$	−1.2	$\frac{19}{10}$	0.75	−1.5	$-\frac{6}{5}$	$-\frac{3}{5}$	1.2	0.3
1.5	1.9	−0.75	−0.4	$\frac{3}{4}$	$-\frac{5}{4}$	−1.9	$\frac{2}{5}$	$-\frac{3}{20}$	$-\frac{19}{10}$
$\frac{6}{5}$	$-\frac{3}{10}$	1.6	$-\frac{2}{5}$	0.6	0.15	$\frac{3}{2}$	−1.25	0.4	$-\frac{8}{5}$

To Play:

- Play with a partner.
- Deal 20 cards to each player face-down.
- Each player turns one card face-up. The player with the greater number wins. The winner collects both cards and places them at the bottom of his or her cards.
- Suppose there is a tie. Each player lays three cards face-down, then a new card face-up. The player with the greater of these new cards wins. The winner collects all ten cards and places them at the bottom of his or her cards.
- Continue playing until one player has all the cards. This player wins the game.

What Is Your Answer?

3. IN YOUR OWN WORDS How can you use a number line to order rational numbers? Give an example.

The numbers are in order from least to greatest. Fill in the blank spaces with rational numbers.

4. $-\frac{1}{2}$, ☐, $\frac{1}{3}$, ☐, $\frac{7}{5}$, ☐

5. $-\frac{5}{2}$, ☐, -1.9, ☐, $-\frac{2}{3}$, ☐

6. $-\frac{1}{3}$, ☐, -0.1, ☐, $\frac{4}{5}$, ☐

7. -3.4, ☐, -1.5, ☐, 2.2, ☐

Name ______________________________ Date __________

3.1 Practice
For use after Lesson 3.1

Write the rational number as a decimal.

1. $-\frac{9}{10}$ **2.** $-4\frac{2}{3}$ **3.** $1\frac{7}{16}$

Write the decimal as a fraction or mixed number in simplest form.

4. -0.84 **5.** 5.22 **6.** -1.716

Order the numbers from least to greatest.

7. $\frac{1}{5}, 0.1, -\frac{1}{2}, -0.25, 0.3$ **8.** $-1.6, \frac{5}{2}, -\frac{7}{8}, 0.9, -\frac{6}{5}$ **9.** $-\frac{2}{3}, \frac{5}{9}, 0.5, -1.3, -\frac{10}{3}$

10. Relative to ground level, a black garden ant digs $-20\frac{7}{9}$ feet and a red harvester ant digs $-20\frac{39}{50}$ feet. Which ant is closer to ground level?

11. The table shows the position of each runner relative to when the first place finisher crossed the finish line. Who finished in second place? Who finished in fifth place?

Runner	A	B	C	D	E	F
Meters	-1.264	$-\frac{5}{4}$	-1.015	-0.480	$-\frac{14}{25}$	$-\frac{13}{8}$

Name____ Date__________

3.2 Adding and Subtracting Rational Numbers

For use with Activity 3.2

Essential Question How does adding and subtracting rational numbers compare with adding and subtracting integers?

1 ACTIVITY: Adding and Subtracting Rational Numbers

Work with a partner. Use a number line to find the sum or difference.

a. $2.7 + (-3.4)$

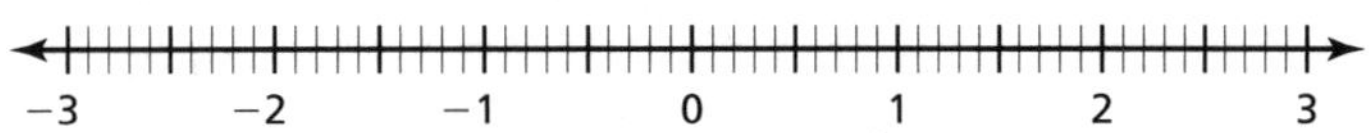

b. $\frac{3}{10} + \left(-\frac{9}{10}\right)$

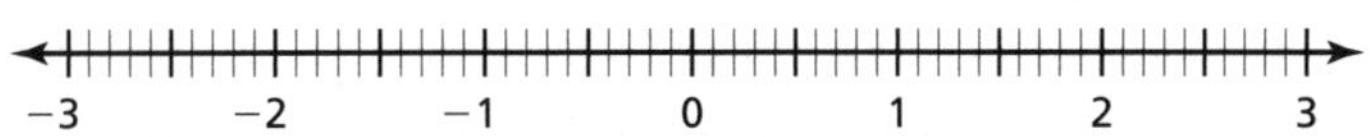

c. $-\frac{6}{10} - 1\frac{3}{10}$

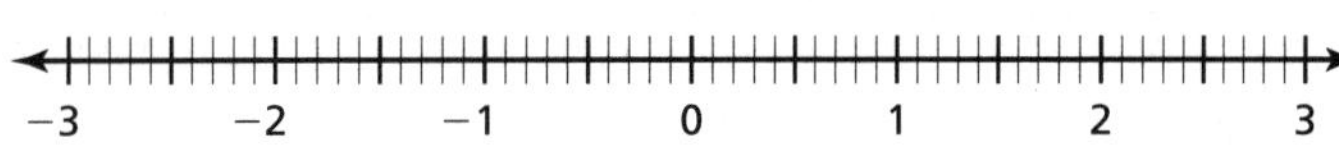

d. $1.3 + (-3.4)$

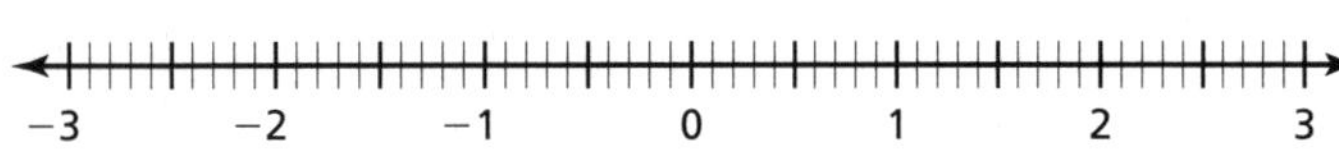

e. $-1.9 - 0.8$

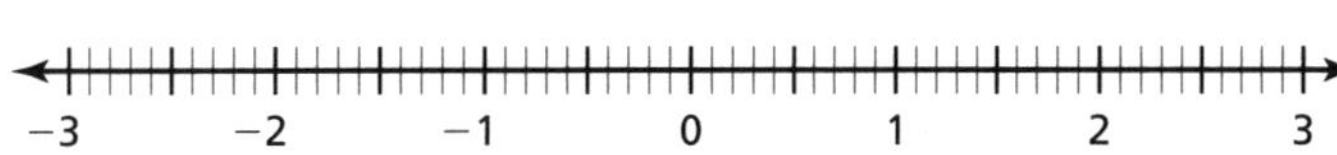

Name ______________________________ Date __________

3.2 Adding and Subtracting Rational Numbers (continued)

2 ACTIVITY: Adding and Subtracting Rational Numbers

Work with a partner. Write the numerical expression shown on the number line. Then find the sum or difference.

a.

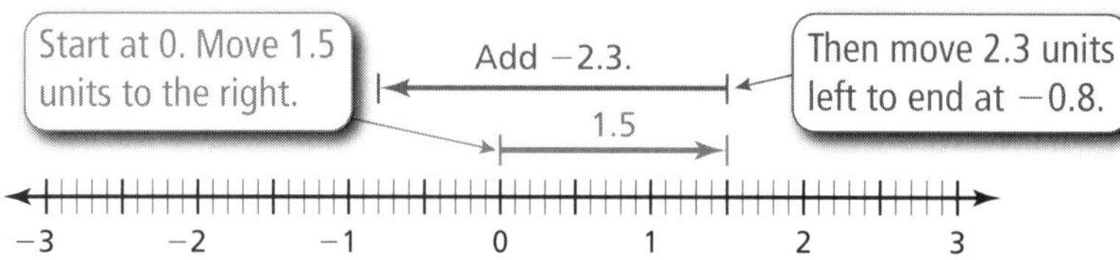

b.

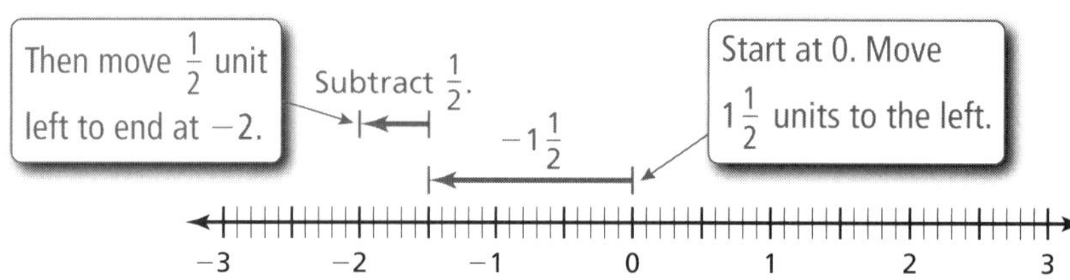

3 ACTIVITY: Financial Literacy

Work with a partner. The table shows the balance in a checkbook.

- Deposits and interest are amounts added to the account.
- Amounts shown in parentheses are taken from the account.

Date	Check #	Transaction	Amount	Balance
--	--	Previous Balance	--	100.00
1/02/2009	124	Groceries	(34.57)	
1/06/2009		Check deposit	875.50	
1/11/2009		ATM withdrawal	(40.00)	
1/14/2009	125	Electric company	(78.43)	
1/17/2009		Music store	(10.55)	
1/18/2009	126	Shoes	(47.21)	
1/20/2009		Check deposit	125.00	
1/21/2009		Interest	2.12	
1/22/2009	127	Cell phone	(59.99)	

3.2 Adding and Subtracting Rational Numbers (continued)

You can find the balance in the second row two different ways.

$100.00 - 34.57 = 65.43$ Subtract 34.57 from 100.00.

$100.00 + (-34.57) = 65.43$ Add –34.57 to 100.00.

a. Complete the balance column of the table on the previous page.

b. How did you find the balance in the tenth row?

c. Use a different way to find the balance in part (b).

What Is Your Answer?

4. IN YOUR OWN WORDS How does adding and subtracting rational numbers compare with adding and subtracting integers? Give an example.

PUZZLE **Find a path through the table so that the numbers add up to the sum. You can move horizontally or vertically.**

5. Sum: $\frac{3}{4}$

Start →

$\frac{1}{2}$	$\frac{2}{3}$	$-\frac{5}{7}$
$-\frac{1}{8}$	$-\frac{3}{4}$	$\frac{1}{3}$

← End

6. Sum: –0.07

Start →

2.43	1.75	−0.98
−1.09	3.47	−4.88

← End

Name ______________________________ Date __________

3.2 Practice

For use after Lesson 3.2

Add or subtract. Write fractions in simplest form.

1. $-\frac{4}{5} + \frac{3}{20}$

2. $-8 + \left(-\frac{6}{7}\right)$

3. $1\frac{2}{15} + \left(-3\frac{1}{2}\right)$

4. $-\frac{1}{6} - \frac{5}{12}$

5. $\frac{9}{10} - 3$

6. $5\frac{3}{4} - \left(-4\frac{5}{6}\right)$

7. $0.46 + (-0.642)$

8. $0.13 - 5.7$

9. $-2.57 - (-3.48)$

10. Tubs of ice cream are delivered to a store at a temperature of 36.7°F. The ice cream is stored in a –40°F freezer. When a tub is brought out of the freezer, its temperature is 22.2°F. Write the difference between the temperatures of the ice cream after the ice cream is in the freezer and before it is in the freezer.

11. Before a race, you start $4\frac{5}{8}$ feet behind your friend. At the halfway point, you are $3\frac{2}{3}$ feet ahead of your friend. What is the change in distance between you and your friend from the beginning of the race?

Name____ Date__________

3.3 Multiplying and Dividing Rational Numbers

For use with Activity 3.3

Essential Question How can you use operations with rational numbers in a story?

1 EXAMPLE: Writing a Story

Write a story that uses addition, subtraction, multiplication, or division of rational numbers. Draw pictures for your story.

There are many possible stories. Here is an example.

Lauryn decides to earn some extra money. She sets up a lemonade stand. To get customers, she uses big plastic glasses and makes a sign saying "All you can drink for 50¢!"

Lauryn can see that her daily profit is negative. But, she decides to keep trying. After one week, she has the same profit each day.

Sunday	Monday	Tuesday	Wednesday	Thursday	Friday	Saturday
-$5.75	-$5.75	-$5.75	-$5.75	-$5.75	-$5.75	-$5.75

Lauryn is frustrated. Her daily profit for the first week is

$$7(-5.75) = (-5.75) + (-5.75) + (-5.75) + (-5.75) + (-5.75) + (-5.75) + (-5.75)$$
$$= -40.25.$$

She realizes that she has too many customers who are drinking a second and even a third glass of lemonade. So, she decides to try a new strategy. Soon, she has a customer. He buys a glass of lemonade and drinks it.

He hands the empty glass to Lauryn and says "*That was great. I'll have another glass.*" Today, Lauryn says "*That will be 50¢ more, please.*" The man says "*But, you only gave me one glass and the sign says 'All you can drink for 50¢!'*" Lauryn replies, "*One glass IS all you can drink for 50¢.*"

With her new sales strategy, Lauryn starts making a profit of $8.25 per day. Her profit for the second week is

$$7(8.25) = (8.25) + (8.25) + (8.25) + (8.25) + (8.25) + (8.25) + (8.25) = 57.75.$$

Her profit for the two weeks is $-40.25 + 57.75 = \$17.50$. So, Lauryn has made some money. She decides that she is on the right track.

3.3 Multiplying and Dividing Rational Numbers (continued)

2 ACTIVITY: Writing a Story

Work with a partner. Write a story that uses addition, subtraction, multiplication, or division of rational numbers.

- At least one of the numbers in the story has to be negative and *not* an integer.
- Draw pictures to help illustrate what is happening in the story.
- Include the solution of the problem in the story.

If you are having trouble thinking of a story, here are some common uses of negative numbers.

- A profit of –\$15 is a loss of \$15.
- An elevation of –100 feet is a depth of 100 feet below sea level.
- A gain of –5 yards in football is a loss of 5 yards.
- A score of –4 in golf is 4 strokes under par.
- A balance of –\$25 in your checking account means the account is overdrawn by \$25.

What Is Your Answer?

3. **IN YOUR OWN WORDS** How can you use operations with rational numbers in a story? You already used rational numbers in your story. Describe another use of a negative rational number in a story.

PUZZLE Read the cartoon. Fill in the blanks using 4's or 8's to make the equation true.

"Dear Mom, I'm in a hurry. To save time I won't be typing any 4's or 8's."

4. $\left(-\dfrac{1}{\square}\right)+\left(-\dfrac{1}{\square}\right)=-\dfrac{1}{\square}$

5. $\left(-\dfrac{1}{\square}\right)\times\left(-\dfrac{1}{\square}\right)=\dfrac{1}{6\square}$

6. $1.\square \times \left(-0.\square\right) = -1.\square\square$

7. $\left(-\dfrac{3}{\square}\right)\div\left(\dfrac{3}{\square}\right)=-\dfrac{1}{2}$

8. $-4.\square \div 2 = -2.\square$

Name ______________________________ Date __________

3.3 Practice

For use after Lesson 3.3

Multiply or divide. Write fractions in simplest form.

1. $-\frac{8}{9}\left(-\frac{18}{25}\right)$

2. $-4\left(\frac{9}{16}\right)$

3. $-3\frac{3}{7} \times 2\frac{1}{2}$

4. $-\frac{2}{3} \div \frac{5}{9}$

5. $\frac{7}{13} \div (-2)$

6. $-5\frac{5}{8} \div \left(-4\frac{7}{12}\right)$

7. $-1.39 \times (-6.8)$

8. $-10 \div 0.22$

9. $-12.166 \div (-1.54)$

10. In a game of tug of war, your team changes $-1\frac{3}{10}$ feet in position every 10 seconds. What is your change in position after 30 seconds?

11. The table shows the change of gas prices over a month's time. What is the mean change?

Week	Change
1	−\$0.06
2	+\$0.10
3	−\$0.08
4	+\$0.02

Name____________________ Date__________

3.4 Solving Equations Using Addition or Subtraction

For use with Activity 3.4

Essential Question How can you use inverse operations to solve an equation?

1 EXAMPLE: Using Addition to Solve an Equation

Use algebra tiles to model and solve $x - 3 = -4$.

Model the equation $x - 3 = -4$.
Draw a sketch of your tiles.

To get the green tile by itself, remove the red tiles on the left side by adding three yellow tiles to each side.

Remove the "zero pairs" from each side.
Draw a sketch of the remaining tiles.

The remaining tiles show the value of x.

$x =$ __________

2 EXAMPLE: Using Addition to Solve an Equation

Use algebra tiles to model and solve $-5 = n + 2$.

________ $= n$ or $n =$ ________

Name ______________________________ Date __________

3.4 Solving Equations Using Addition or Subtraction (continued)

3 ACTIVITY: Solving Equations Using Algebra Tiles

Work with a partner. Use algebra tiles to model and solve the equation.

a. $y + 10 = -5$

b. $p - 7 = -3$

c. $-15 = t - 5$

d. $8 = 12 + z$

4 ACTIVITY: Writing and Solving Equations

Work with a partner. Write an equation shown by the algebra tiles. Then solve.

a.

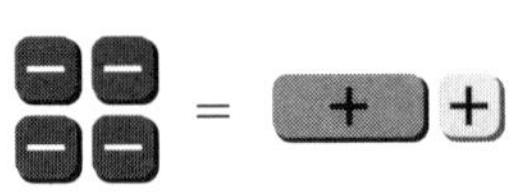

b.

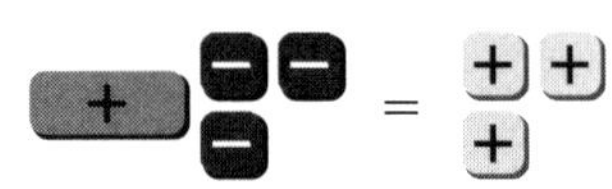

c.

d.

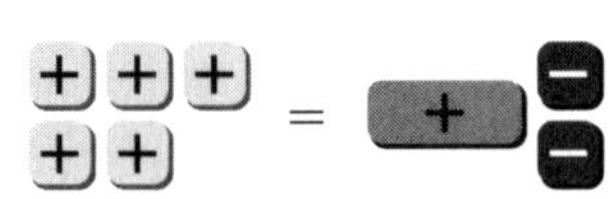

Name____ Date__________

3.4 **Solving Equations Using Addition or Subtraction (continued)**

What Is Your Answer?

5. Decide whether the statement is *true* or *false*. Explain your reasoning.

a. In an equation, any letter can be used as a variable. ______________

b. The goal in solving an equation is to get the variable by itself. ______________

c. In the solution, the variable always has to be on the left side of the equal sign. ______________

d. If you add a number to one side, you should add it to the other side. ______________

6. IN YOUR OWN WORDS How can you use inverse operations to solve an equation without algebra tiles? Give two examples.

7. What makes the cartoon funny?

"Dear Sir: Yesterday you said $x = 2$. Today you are saying $x = 3$. Please make up your mind."

8. The word *variable* comes from the word *vary*. For example, the temperature in Maine varies a lot from winter to summer. Write two other English sentences that use the word *vary*.

Name ______________________________ Date __________

3.4 Practice

For use after Lesson 3.4

Solve the equation. Check your solution.

1. $y + 12 = -26$

2. $15 + c = -12$

3. $-16 = d + 21$

4. $n + 12.8 = -0.3$

5. $1\frac{1}{8} = g - 4\frac{2}{5}$

6. $-5.47 + k = -14.19$

Write the word sentence as an equation. Then solve.

7. 42 less than x is -50.

8. 32 is the sum of a number z and 9.

9. A clothing company makes a profit of $2.3 million. This is $4.1 million more than last year. What was the profit last year?

10. A drop on a wooden roller coaster is $-98\frac{1}{2}$ feet. A drop on a steel roller coaster is $100\frac{1}{4}$ feet lower than the drop on the wooden roller coaster. What is the drop on the steel roller coaster?

Name____ Date__________

3.5 Solving Equations Using Multiplication or Division

For use with Activity 3.5

Essential Question How can you use multiplication or division to solve an equation?

1 ACTIVITY: Using Division to Solve an Equation

Work with a partner. Use algebra tiles to model and solve the equation.

a. $3x = -12$

Model the equation $3x = -12$. Draw a sketch of your tiles.

Your goal is to get one green tile by itself. Because there are three green tiles, divide the red tiles into three equal groups.

Keep one of the groups. This shows the value of x. Draw a sketch of the remaining tiles.

$x =$ __________.

b. $2k = -8$

c. $-15 = 3t$

d. $-20 = 5m$

e. $4h = -16$

Name ______________________________ Date ________

2 ACTIVITY: Writing and Solving Equations

Work with a partner. Write an equation shown by the algebra tiles. Then solve.

a. 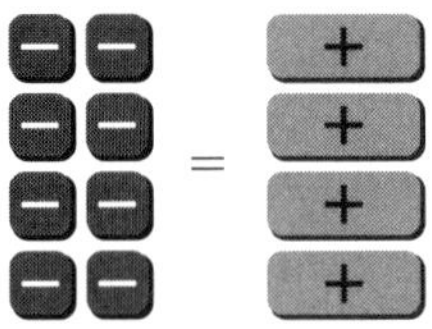

b. 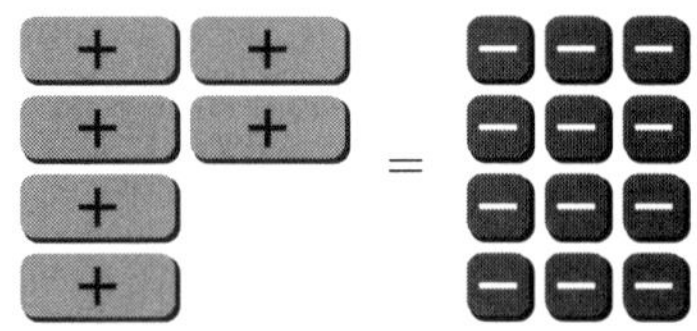

c. 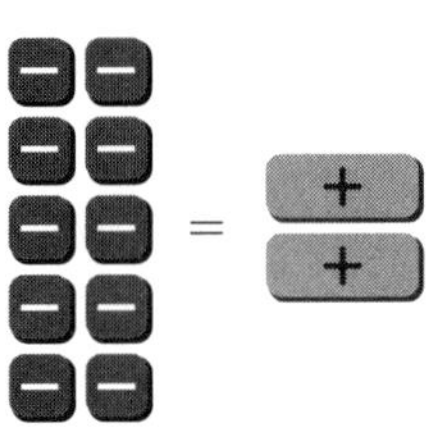

d. 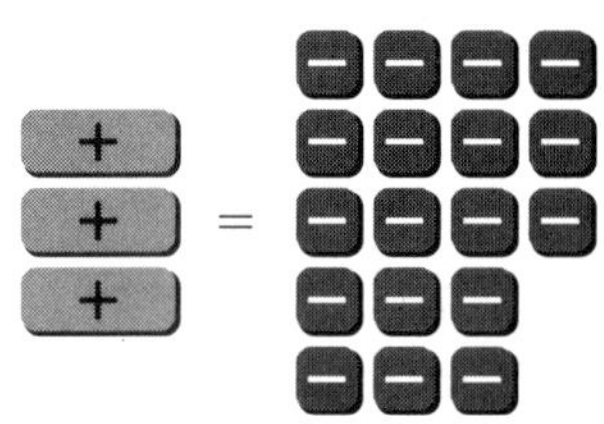

3 ACTIVITY: The Game of Math Card War

Preparation:

- Cut index cards to make 40 playing cards.
- Write each equation in the table on the next page on the card.

$-4x = -12$	$x - 1 = 1$	$x - 3 = 1$	$2x = -10$	$-9 = 9x$
$3 + x = -2$	$x = -2$	$-3x = -3$	$\frac{x}{-2} = -2$	$x = -6$
$6x = -36$	$-3x = -9$	$-7x = -14$	$x - 2 = 1$	$-1 = x + 5$
$x = -1$	$9x = -27$	$\frac{x}{3} = -1$	$-8 = -2x$	$x = 3$
$-7 = -1 + x$	$x = -5$	$-10 = 10x$	$x = -4$	$-2 = -3 + x$
$-20 = 10x$	$x + 9 = 8$	$-16 = 8x$	$x = 2$	$x + 13 = 11$
$x = -3$	$-8 = 2x$	$x = 1$	$\frac{x}{2} = -2$	$-4 + x = -2$
$\frac{x}{5} = -1$	$-6 = x - 3$	$x = 4$	$x + 6 = 2$	$x - 5 = -4$

To Play:

- Play with a partner. Deal 20 cards to each player face-down.
- Each player turns one card face-up. The player with the greater solution wins. The winner collects both cards and places them at the bottom of his or her cards.
- Suppose there is a tie. Each player lays three cards face-down, then a new card face-up. The player with the greater solution of these new cards wins. The winner collects all ten cards, and places them at the bottom of his or her cards.
- Continue playing until one player has all the cards. This player wins the game.

What Is Your Answer?

4. IN YOUR OWN WORDS How can you use multiplication or division to solve an equation without using algebra tiles? Give two examples.

Name ____ Date __________

3.5 Practice

For use after Lesson 3.5

Solve the equation. Check your solution.

1. $\frac{d}{5} = -6$

2. $8x = -6$

3. $-15 = \frac{z}{-2}$

4. $3.2n = -0.8$

5. $-\frac{3}{10}h = 15$

6. $-1.1k = -1.21$

Write the word sentence as an equation. Then solve.

7. A number divided by –8 is 7.

8. The product of –12 and a number is 60.

9. You earn $0.85 for every cup of hot chocolate you sell. How many cups do you need to sell to earn $55.25?

10. The price of a satellite radio at Store A is $\frac{29}{30}$ the price at Store B. The price at Store A is $150.80.

a. Write and solve an equation to find the price of the satellite radio at Store B.

b. How much do you save by buying the satellite radio at Store A?

Name____ Date__________

3.6 Solving Two-Step Equations
For use with Activity 3.6

Essential Question In a two-step equation, which step should you do first?

1 EXAMPLE: Solving a Two-Step Equation

Use algebra tiles to model and solve $2x - 3 = -5$.

Model the equation $2x - 3 = -5$.
Draw a sketch of your tiles.

Remove the three red tiles on the left side by adding __________ yellow tiles to each side.

Because there are two green tiles, divide the red tiles into __________ equal groups.

Keep one of the groups. This shows the value of x.
Draw a sketch of the remaining tiles.

$x =$ ________.

2 EXAMPLE: The Math Behind the Tiles

Solve $2x - 3 = -5$ without using algebra tiles. Describe each step. Which step is first, adding 3 to each side or dividing each side by 2?

$x =$ _____. The first step is ______________________________.

3 ACTIVITY: Solving Equations Using Algebra Tiles

Work with a partner.

- **Write an equation shown by the algebra tiles.**
- **Use algebra tiles to model and solve the equation.**
- **Check your answer by solving the equation without using algebra tiles.**

a. 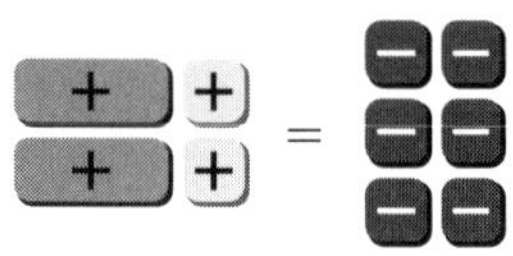

b. 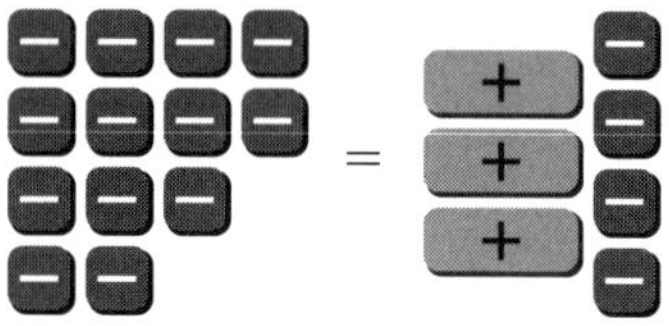

4 ACTIVITY: Working Backwards

Work with a partner.

a. Your friend pauses a video game to get a drink. You continue the game. You double the score by saving a princess. Then you lose 75 points because you do not collect the treasure. You finish the game with –25 points. How many points did you start with?

One way to solve the problem is to work backwards. To do this, start with the end result and retrace the events.

You started the game with ____________ points.

b. You triple your account balance by making a deposit. Then you withdraw \$127.32 to buy groceries. Your account is now overdrawn by \$10.56. By working backwards, find your account balance before you made the deposit.

What Is Your Answer?

5. IN YOUR OWN WORDS In a two-step equation, which step should you do first? Give four examples.

6. Solve the equation $2x - 75 = -25$. How do your steps compare with the strategy of working backwards in Activity 4?

Name ______________________________ Date __________

3.6 Practice

For use after Lesson 3.6

Solve the equation. Check your solution.

1. $3a - 5 = -14$

2. $10 = -2c + 22$

3. $18 = -5b - 17$

4. $-12 = -8z + 12$

5. $1.3n - 0.03 = -9$

6. $-\frac{5}{11}h + \frac{7}{9} = \frac{2}{9}$

7. It costs \$34.95 to rent a jet ski for four hours plus \$15.75 for each additional hour. You have \$100. Can you rent the jet ski for 8 hours? Explain.

8. The length of a rectangle is 3 meters less than twice its width.

a. Write an equation to find the length of the rectangle.

b. The length of the rectangle is 11 meters. What is the width of the rectangle?

Name______________________________ Date__________

3.7 Solving Inequalities Using Addition or Subtraction

For use with Activity 3.7

Essential Question How can you use an inequality to describe a real-life situation?

1 ACTIVITY: Writing an Inequality

Work with a partner. In 3 years, your friend will still not be old enough to apply for a driver's license.

a. Which of the following represents your friend's situation? What does x represent? Explain your reasoning.

$x + 3 < 16$	$x + 3 \le 16$	$x + 3 > 16$	$x + 3 \ge 16$

b. Graph the possible ages of your friend on a number line.

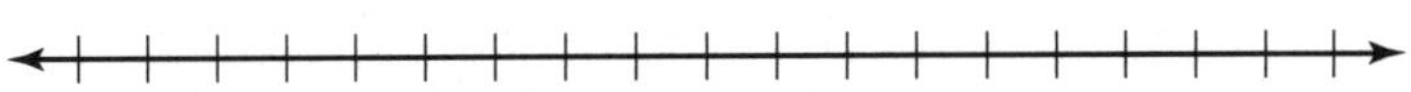

2 ACTIVITY: The Triangle Inequality

Work with a partner. Draw different triangles whose sides have lengths 10 cm, 6 cm, and x cm.

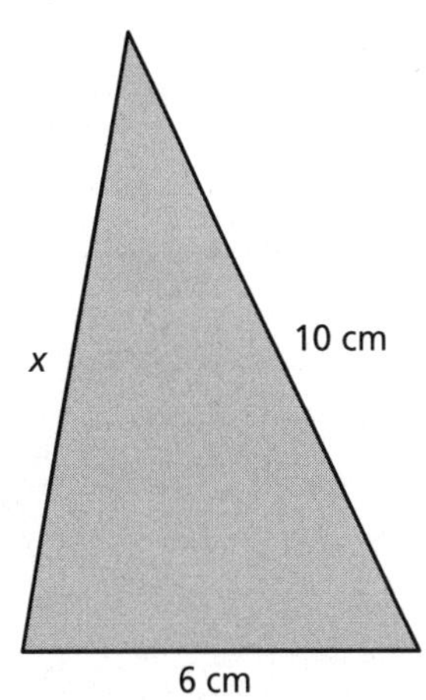

a. Which of the following describes how *small* x can be?

$6 + x < 10$	$6 + x \le 10$
$6 + x > 10$	$6 + x \ge 10$

b. Which of the following describes how *large* x can be?

$x - 6 < 10$	$x - 6 \le 10$
$x - 6 > 10$	$x - 6 \ge 10$

Name ______________________________ Date __________

3.7 Solving Inequalities Using Addition or Subtraction (continued)

3 ACTIVITY: Writing an Inequality

Work with a partner. Baby manatees are about 4 feet long at birth. They grow to a maximum length of 13 feet.

a. Which of the following can represent a baby manatee's growth? What does x represent? Explain your reasoning.

$x + 4 < 13$	$x + 4 \le 13$	$x - 4 > 13$	$x - 4 \ge 13$

b. Graph the solution on a number line.

4 ACTIVITY: Puzzles

Work with a partner.

a. Use the clues to find the word that is spelled by ●, ▶, ■, ♦. Assume $A = 1$, $B = 2$, and so on.

●	▶	■	♦
CLUES	CLUES	CLUES	CLUES
$4 + h \le 7$	$9 \le h - 6$	$h - 5 \le 10$	$7 > h - 6$
$h + 1 > 3$	$3 + h < 19$	$h + 12 > 26$	$11 + h \ge 23$

b. Use the cut-out pieces. Rearrange them to make a square without overlapping the pieces.*

Pentagon to Square

Hexagon to Square

*Cut-outs are available in the back of the Record and Practice Journal.

3.7 Solving Inequalities Using Addition or Subtraction (continued)

c. Use exactly four 4's and the operations +, −, ×, and ÷ to write expressions that have values of 0, 1, 2, 3, 4, 5, 6, 7, 8, 9, and 10. For instance, $44 - 44 = 0$.

What Is Your Answer?

5. **IN YOUR OWN WORDS** How can you use an inequality to describe a real-life situation?

6. Write a real-life situation that you can represent with an inequality. Write the inequality. Graph the solution on a number line.

Name ______________________________ Date ________

3.7 Practice
For use after Lesson 3.7

Solve the inequality. Graph the solution.

1. $x - 4 < 8$

2. $16 + p \geq 14$

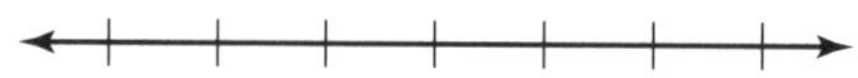

3. $9 > y + \frac{3}{4}$

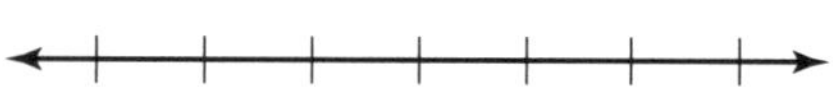

4. $-9.6 \leq z - 2.1$

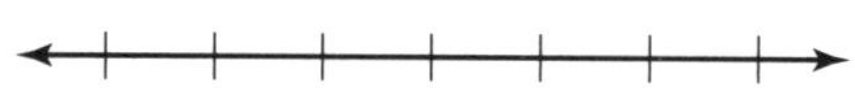

Write and solve an inequality that represents the value of *x*.

5. The perimeter is less than 15 feet.

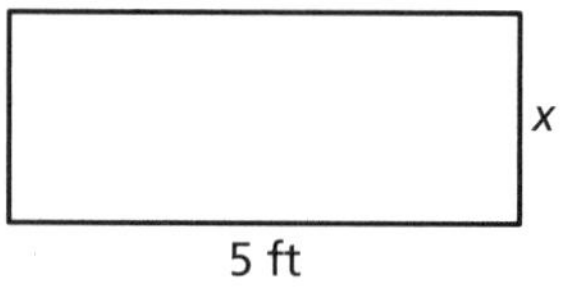

6. The height is greater than the base.

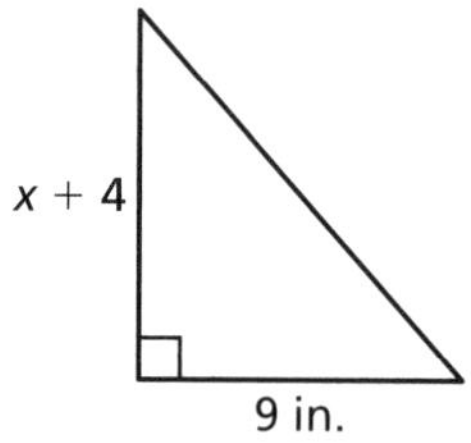

7. Your goal is to sell at least 50 boxes of cookies for your school fundraiser.

a. Write an inequality that represents your goal.

b. You sell 26 boxes. Write and solve a new inequality to represent how many boxes you need to sell to reach your goal.

Name____________________________ Date__________

3.8 Solving Inequalities Using Multiplication or Division

For use with Activity 3.8

Essential Question How can you use multiplication or division to solve an inequality?

1 ACTIVITY: Using a Table to Solve an Inequality

Work with a partner.

- **Complete the table.**
- **Decide which graph represents the solution of the inequality.**
- **Write the solution of the inequality.**

a. $3x \le 6$

x	-1	0	1	2	3	4	5
$3x$							
$3x \overset{?}{\le} 6$							

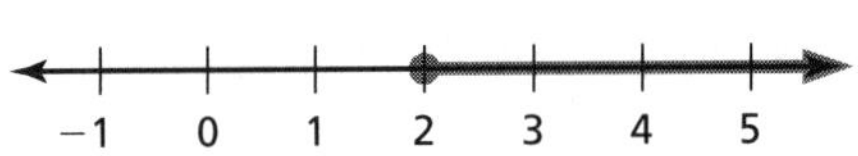

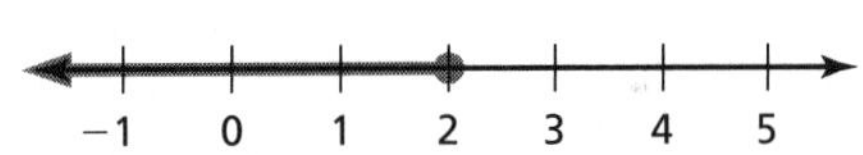

b. $-2x > 4$

x	-5	-4	-3	-2	-1	0	1
$-2x$							
$-2x \overset{?}{>} 4$							

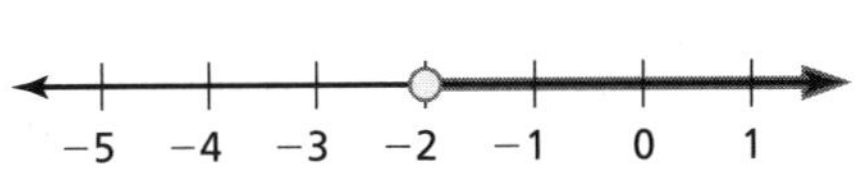

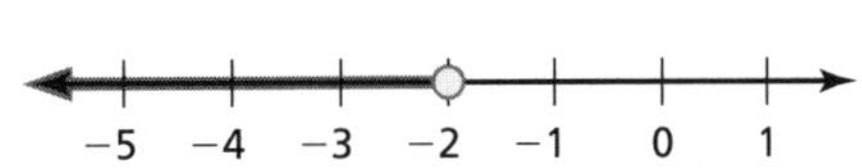

Name ______________________________ Date __________

2 ACTIVITY: Writing a Rule

Work with a partner. Use a table to solve each inequality.

a. $3x > 3$ **b.** $4x \le 4$ **c.** $-2x \ge 6$ **d.** $-5x < 10$

x									
3x									
4x									
−2x									
−5x									

Write a rule that describes how to solve inequalities like those in Activity 1. Then use your rule to solve each of the four inequalities above.

3 ACTIVITY: Using a Table to Solve an Inequality

Work with a partner.

- **Complete the table.**
- **Decide which graph represents the solution of the inequality.**
- **Write the solution of the inequality.**

a. $\frac{x}{2} \ge 1$

x	−1	0	1	2	3	4	5
$\frac{x}{2}$							
$\frac{x}{2} \overset{?}{\ge} 1$							

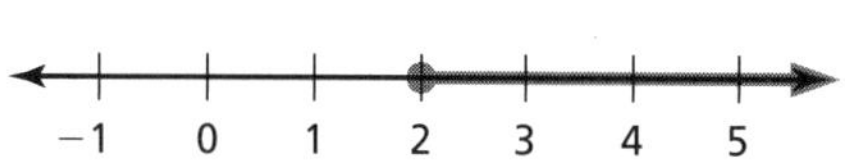

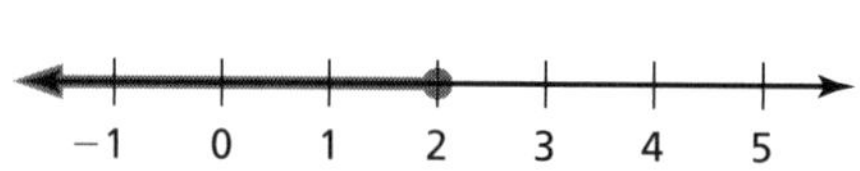

Name________________________ Date__________

3.8 Solving Inequalities Using Multiplication or Division (continued)

b. $\frac{x}{-3} < \frac{2}{3}$

x	-5	-4	-3	-2	-1	0	1
$\frac{x}{-3}$							
$\frac{x}{-3} \overset{?}{<} \frac{2}{3}$							

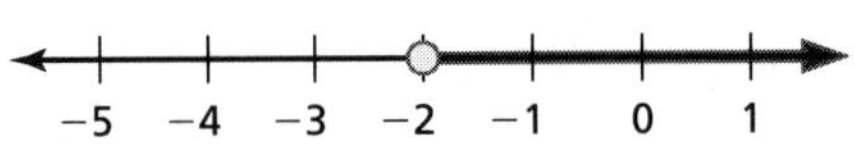

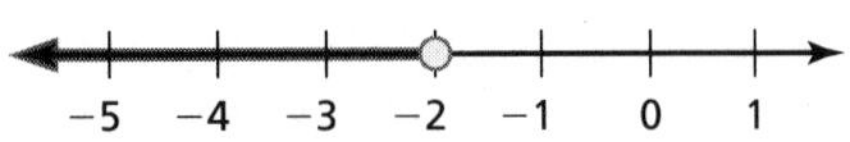

4 ACTIVITY: Writing a Rule

Work with a partner. Use a table to solve each inequality.

a. $\frac{x}{4} \geq 1$ **b.** $\frac{x}{2} < \frac{3}{2}$ **c.** $\frac{x}{-2} > 2$ **d.** $\frac{x}{-5} \leq \frac{1}{5}$

x											
$\frac{x}{4}$											
$\frac{x}{2}$											
$\frac{x}{-2}$											
$\frac{x}{-5}$											

Write a rule that describes how to solve inequalities like those in Activity 3. Then use your rule to solve each of the four inequalities above.

What Is Your Answer?

5. **IN YOUR OWN WORDS** How can you use multiplication or division to solve an inequality?

Name ______________________________ Date __________

3.8 Practice

For use after Lesson 3.8

Solve the inequality. Graph the solution.

1. $5n < 75$

2. $\frac{x}{6} \leq -12$

3. $-15t > -60$

4. $-4q \geq 122$

5. $-8p < \frac{4}{5}$

6. $-9 \geq 2.4m$

7. $-\frac{r}{2} \leq -11$

8. $-\frac{t}{6} > 1.2$

9. $-4 \geq \frac{q}{-0.1}$

10. To win a trivia game, you need at least 60 points. Each question is worth 4 points. Write and solve an inequality that represents the number of questions you need to answer correctly to win the game.

Name______________________________ Date__________

Chapter 4 Fair Game Review

Using the numbers from the In and Out Table, find and state the rule in words.

1.

In	Out
1	4
2	5
3	6
4	7

2.

In	Out
2	6
4	12
6	18
8	24

3.

In	Out
12	2
24	14
36	26
48	38

4.

In	Out
4	2
5	$\frac{5}{2}$
6	3
7	$\frac{7}{2}$

5. The In and Out Table shows the results of buying pretzels from a vending machine. The in column is the amount you put into the machine. The out column is the change you receive back from the machine. Complete the table and state the rule in words.

In	Out
0.65	0
0.70	0.05
0.75	0.10
1.00	

Name ______________________________ Date __________

Chapter 4 Fair Game Review (continued)

Use the graph to write an ordered pair corresponding to the point.

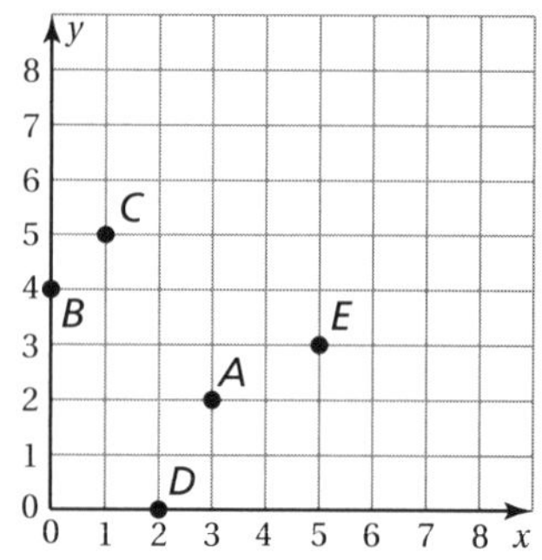

6. Point A

7. Point B

8. Point C

9. Point D

10. Point E

Plot the point on the graph.

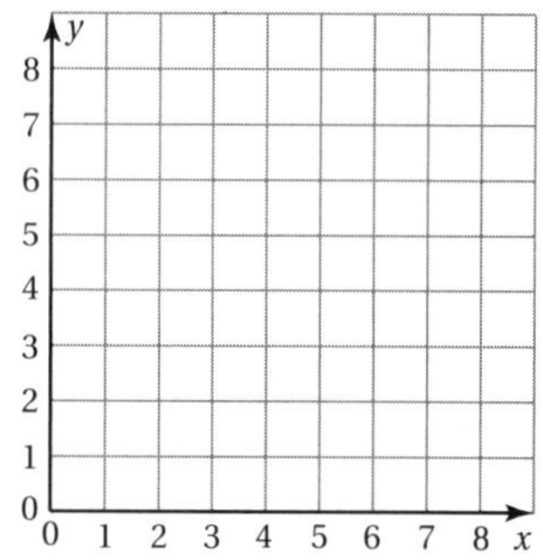

11. $P(2, 5)$

12. $Q(0, 1)$

13. $R(3, 3)$

14. $S(4, 2)$

15. $T(5, 0)$

Name______________________________ Date__________

4.1 Mapping Diagrams

For use with Activity 4.1

Essential Question What is a mapping diagram? How can it be used to represent a function?

1 ACTIVITY: Constructing Mapping Diagrams

Work with a partner. Complete the mapping diagram.

a. Area A

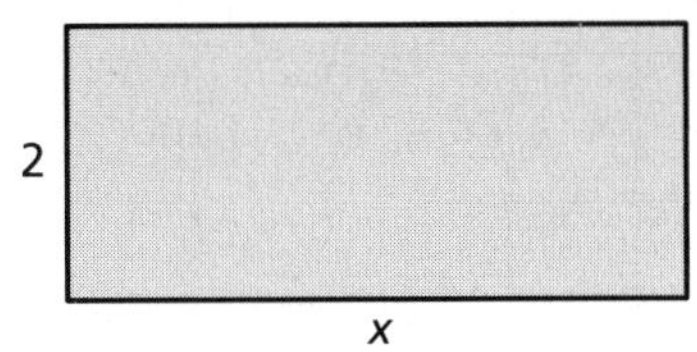

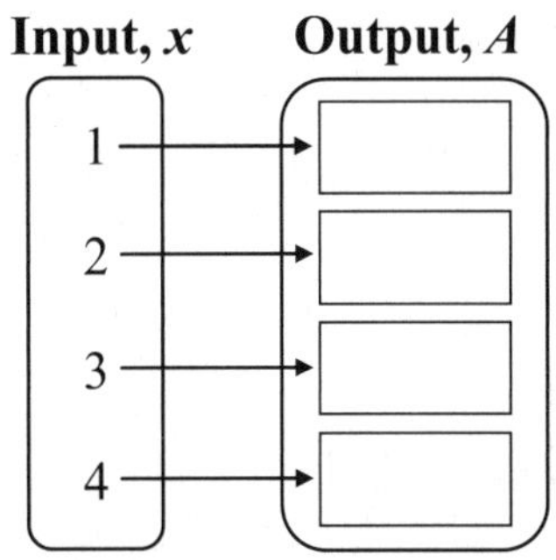

b. Perimeter P

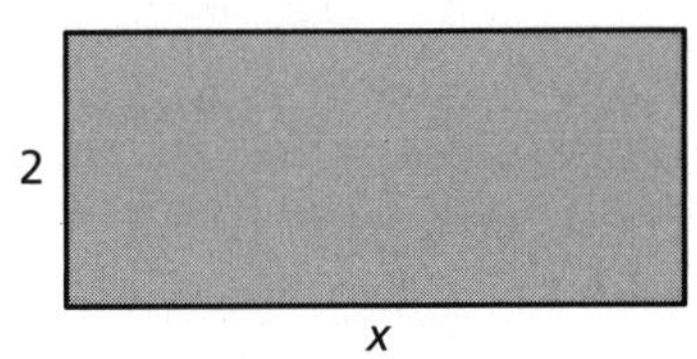

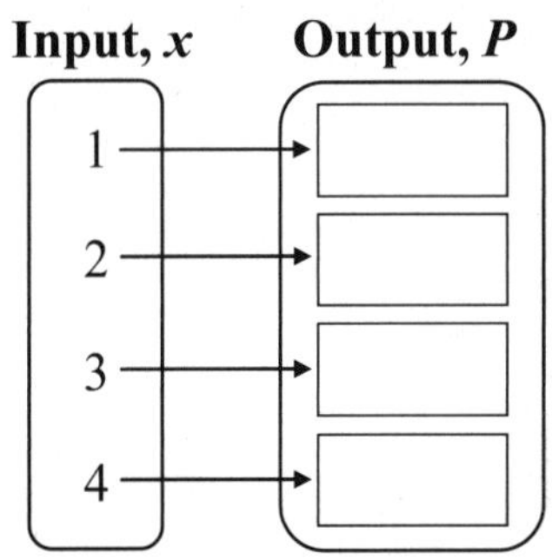

c. Circumference C

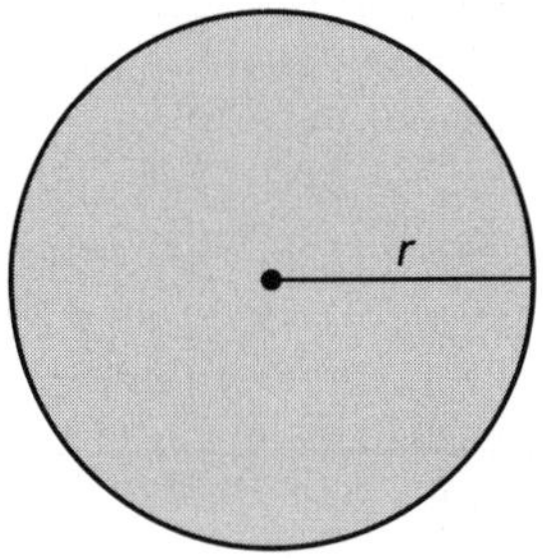

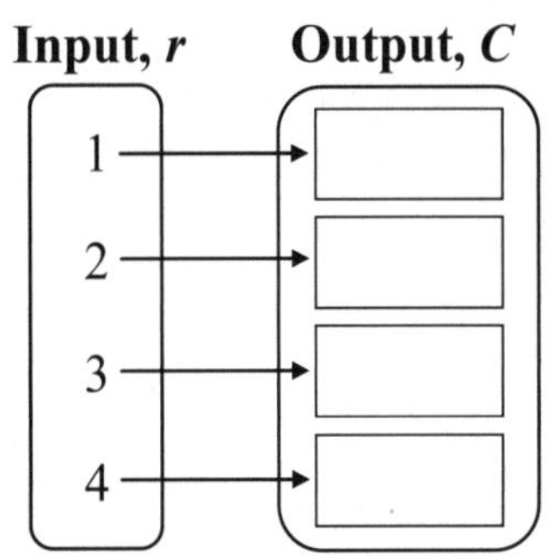

d. Volume V

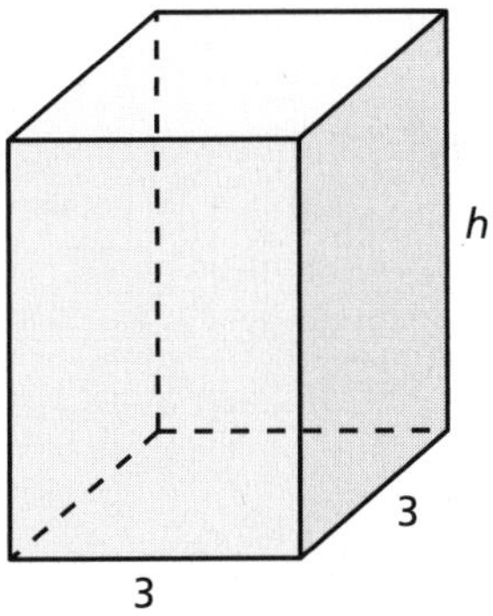

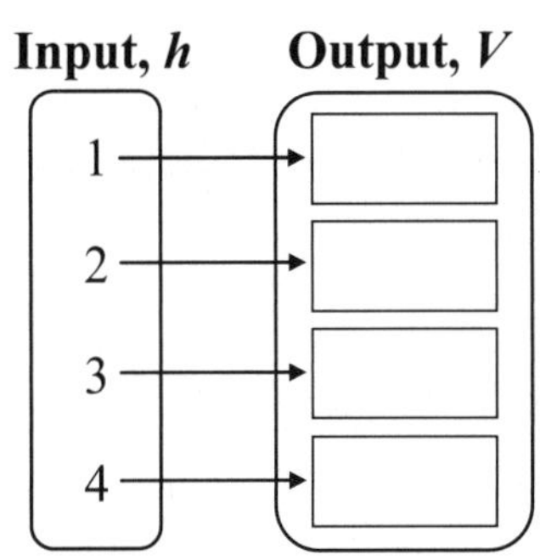

4.1 Mapping Diagrams (continued)

2 ACTIVITY: Interpreting Mapping Diagrams

Work with a partner. Describe the pattern in the mapping diagram. Complete the diagram.

a. **Input, *d*** **Output, *P***

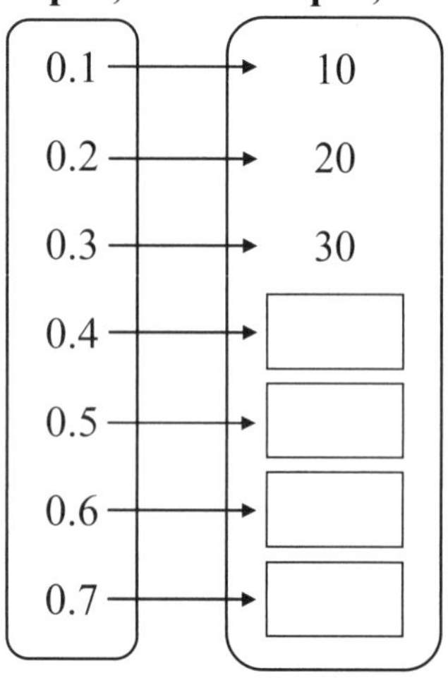

b. **Input, *t*** **Output, *M***

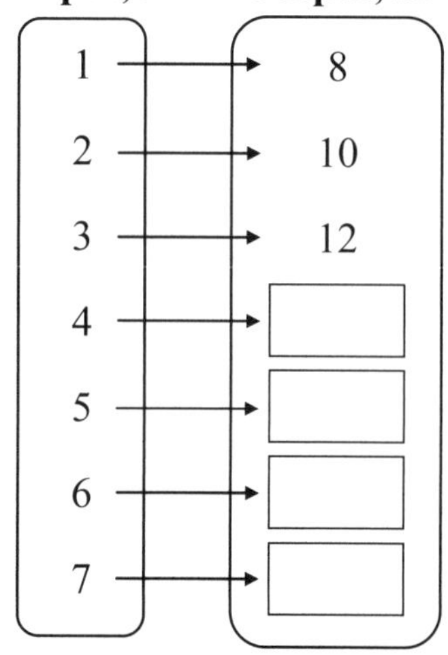

c. **Input, *n*** **Output, *S***

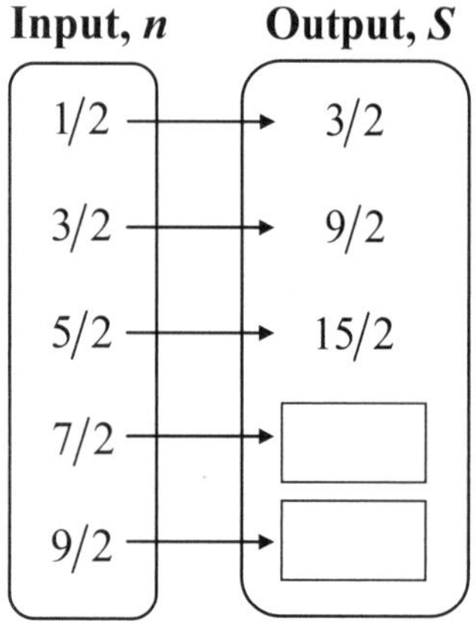

d. **Input, *x*** **Output, *A***

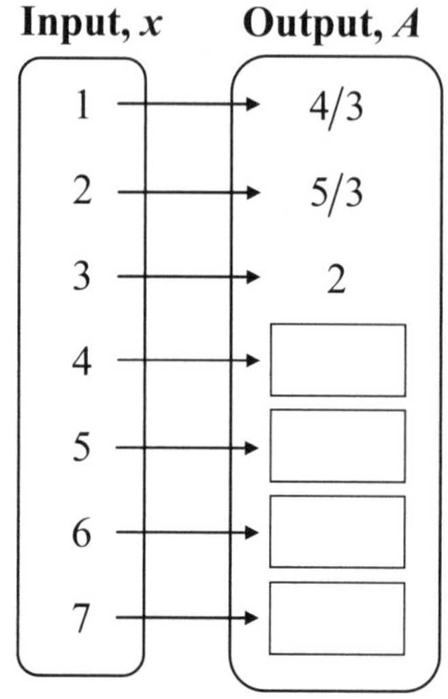

Name____ Date__________

What Is Your Answer?

3. **IN YOUR OWN WORDS** What is a mapping diagram? How can it be used to represent a function?

4. Construct a mapping diagram that represents a function you have studied.

"I made a mapping diagram."

"It shows how I feel about my skateboard with each passing day."

Name ______________________________ Date __________

4.1 Practice

For use after Lesson 4.1

List the ordered pairs shown in the mapping diagram.

1.

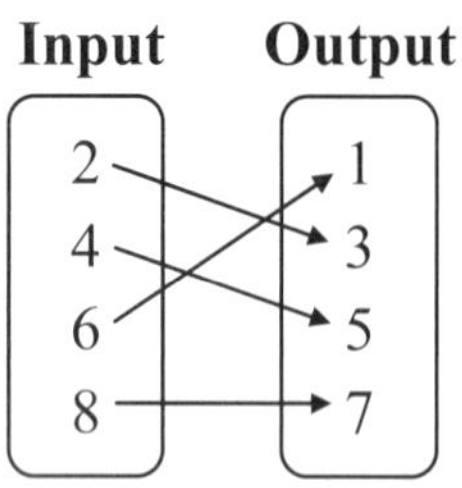

2.

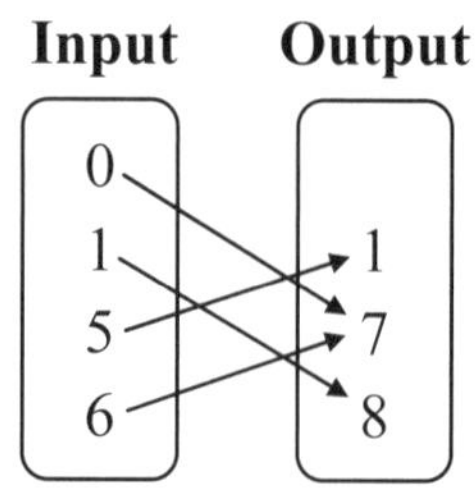

Draw a mapping diagram of the set of ordered pairs.

3. $(1, 7), (3, 4), (4, 8), (7, 2)$

4. $(0, 3), (2, 4), (5, 4), (8, 5)$

5. Draw a mapping diagram for the graph. Then describe the pattern of inputs and outputs.

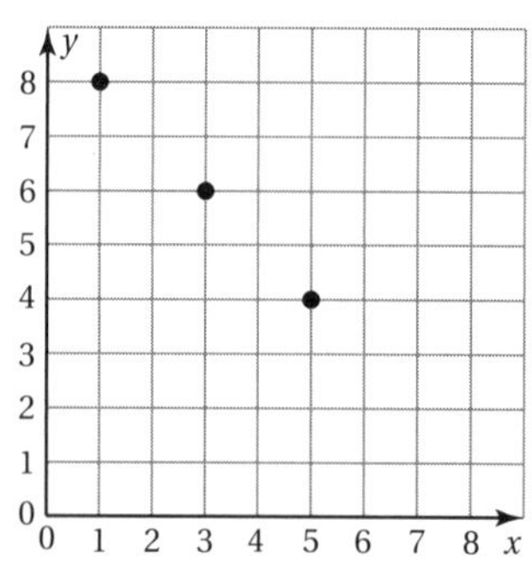

6. The table shows the number of beads needed to make a bracelet. Use the table to draw a mapping diagram.

Bracelet Length (in.)	Number of Beads
6	12
7	14
8	16
9	18

Name________________________________ Date__________

4.2 Functions as Words and Equations

For use with Activity 4.2

Essential Question How can you describe a function with words? How can you describe a function with an equation?

1 ACTIVITY: Describing a Function

Work with a partner. Two mapping diagrams related to the rectangle are shown. Describe each function in words. Then write an equation for each function.

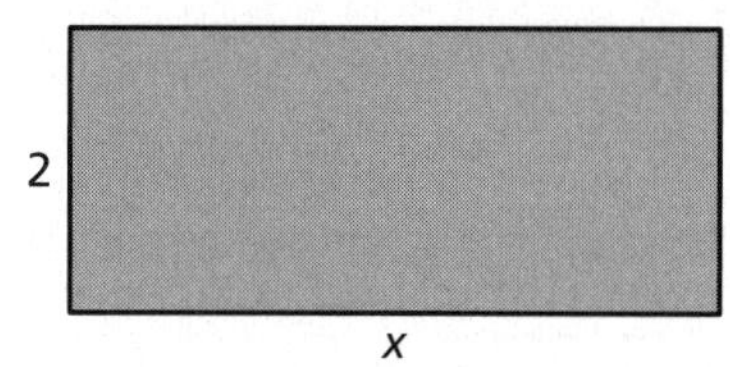

a. Area A

Input, x Output, A

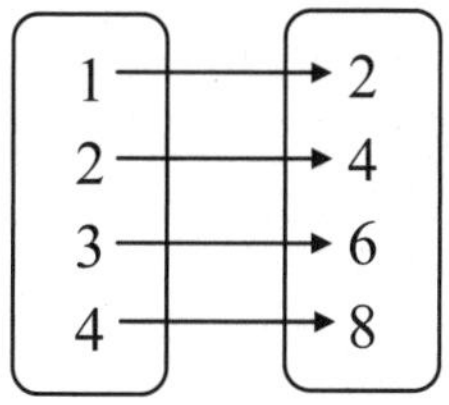

b. Perimeter P

Input, x Output, P

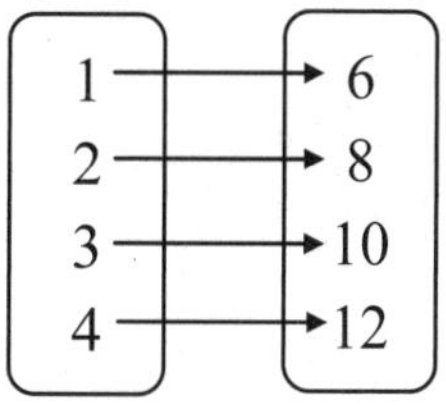

2 ACTIVITY: Describing a Function

Work with a partner. Complete the mapping diagram on the next page for the area of the figure. Then write an equation that describes the function.

a.

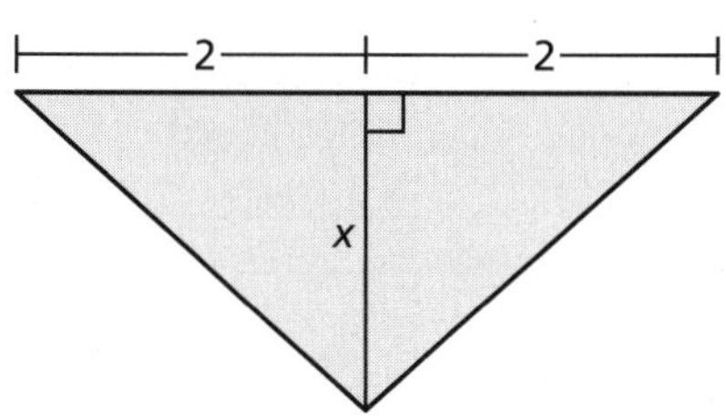

b.

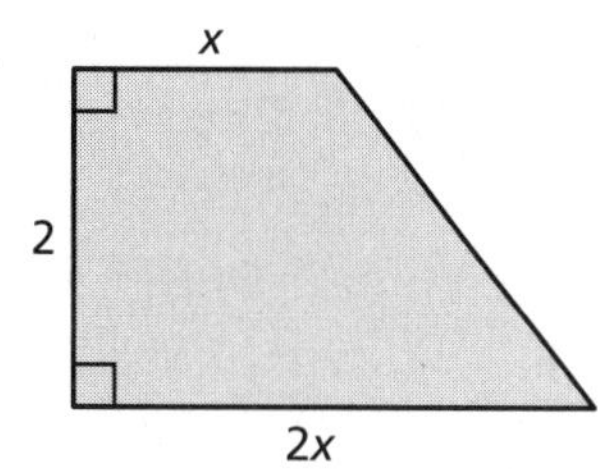

Name ______________________________ Date __________

4.2 Functions as Words and Equations (continued)

Input, x **Output, A**

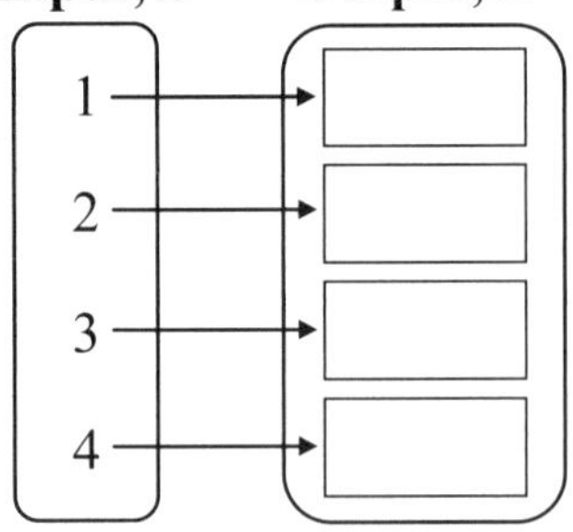

Input, x **Output, A**

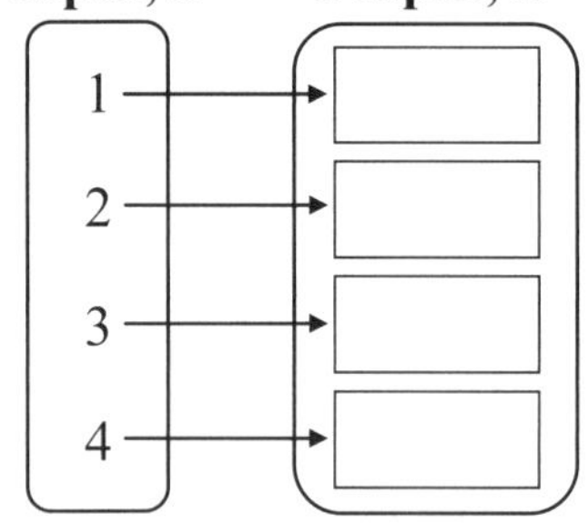

3 ACTIVITY: Describing a Function

Work with a partner. Complete the mapping diagram. Then write an equation that describes the function.

a. Sales tax of 6%

Input, x **Output, T**
Cost (\$) **Sales Tax (\$)**

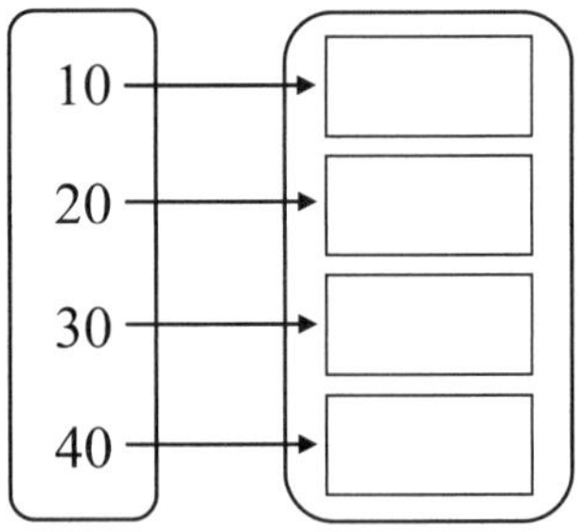

b. Phone bill of \$5 per hour

Input, x **Output, B**
Hours **Bill (\$)**

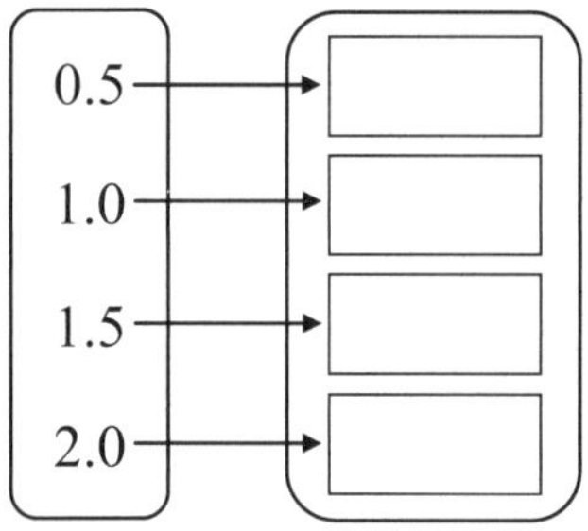

c. Perimeter of a square

Input, s **Output, P**
Side (in.) **Perimeter (in.)**

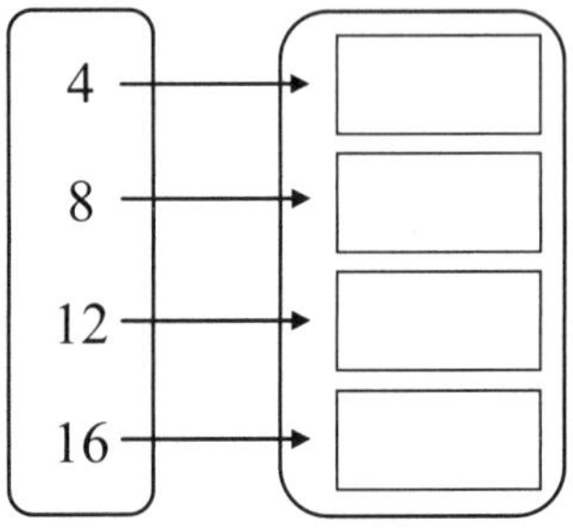

d. Surface area of a cube

Input, s **Output, S**
Side (m) **Surface Area (m^2)**

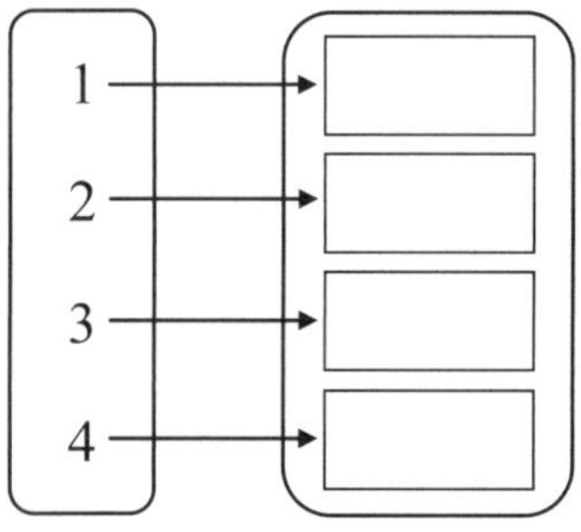

Name______________________________ Date__________

4.2 Functions as Words and Equations (continued)

e. Circumference of a circle

Input, *r* **Output, *C***

Radius (ft) **Circumference (in.)**

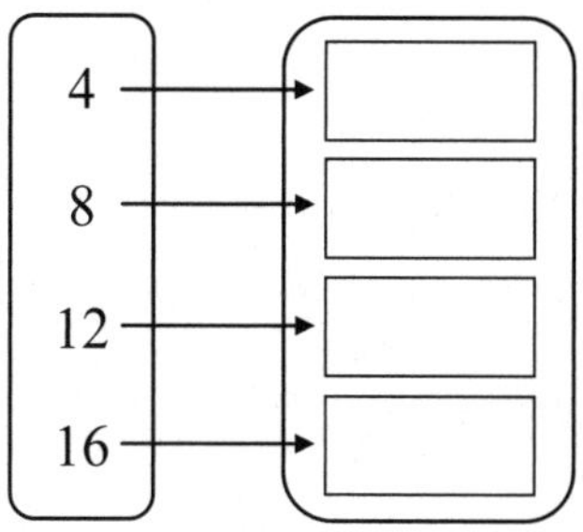

f. Area of a semicircle

Input, *r* **Output, *A***

Radius (cm) **Area (cm^2)**

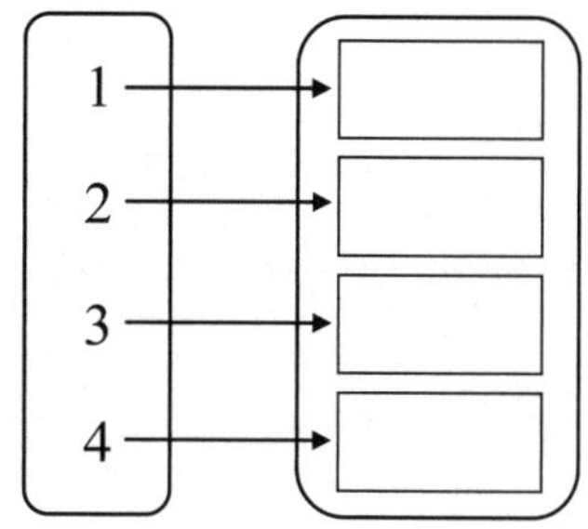

What Is Your Answer?

4. IN YOUR OWN WORDS How can you describe a function with words? How can you describe a function with an equation? Give some examples from lessons you have studied this year.

Name ______________________________ Date __________

4.2 Practice

For use after Lesson 4.2

Write an equation that describes the function.

1. The output is four times the input.

2. The output is eight less than the input.

Find the value of *y* for the given value of *x*.

3. $y = \frac{x}{3}; x = 12$

4. $y = 5x + 9; x = 2$

Tell whether the ordered pair is a solution of the equation.

5. $y = 6x - 7; (2, 5)$

6. $y = \frac{x}{2} + 1; (3, 4)$

7. You set up a hot chocolate stand at a football game. The cost of your supplies is \$75. You charge \$0.50 for each cup of hot chocolate.

a. Write an equation you can use to find the profit *P* for selling *c* cups of hot chocolate.

b. You will *break even* if the cost of your supplies equals your income. How many cups of hot chocolate must you sell to break even?

Name______________________________ Date__________

4.3 Input-Output Tables

For use with Activity 4.3

Essential Question How can you use a table to describe a function?

1 ACTIVITY: Using a Function Table

Work with a partner.

a. Complete the table for the perimeter of the rectangle.

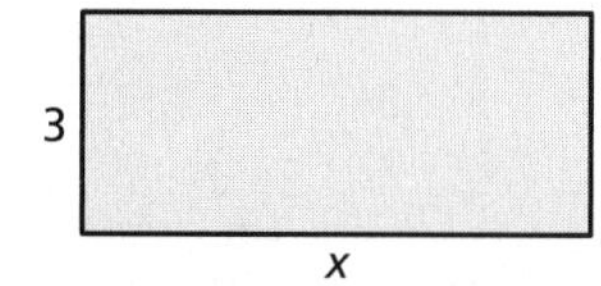

Input, *x*	1	2	3	4	5
Output, *P*					

b. Write an equation that describes the function.

c. Use your equation to find the values of x for which the perimeter is 50.

Name ______________________________ Date ________

4.3 Input-Output Tables (continued)

2 ACTIVITY: Using a Function Table

Work with a partner. Use the strategy shown in Activity 1 to make a table that shows the pattern for the area. Write an equation that describes the function. Then use your equation to find which figure has an area of 81.

1 square unit

a.

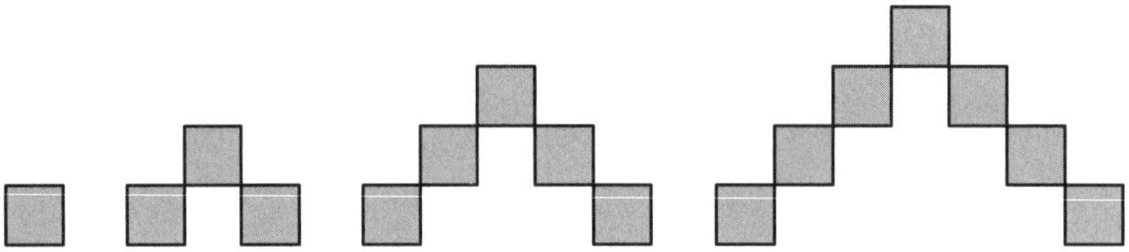

Figure 1 Figure 2 Figure 3 Figure 4

b.

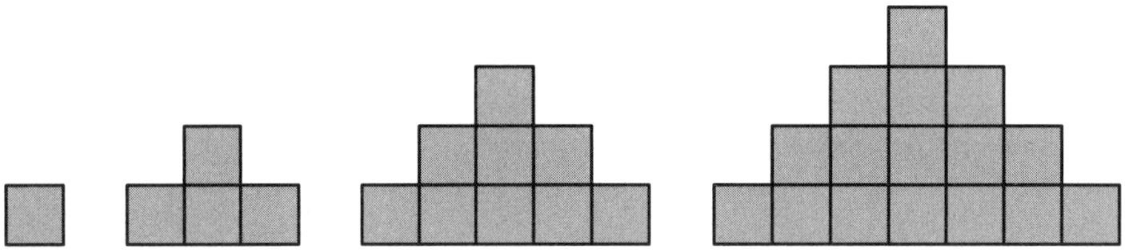

Figure 1 Figure 2 Figure 3 Figure 4

3 ACTIVITY: Making a Function Table

Work with a partner. Complete a sales tax table for each of the four cities.

Madison, WI, 5.50%

Sale, *x*	\$20	\$30	\$40	\$50	\$60
Sales tax, *T*					

Ann Arbor, MI, 6.00%

Sale, *x*	\$20	\$30	\$40	\$50	\$60
Sales tax, *T*					

Name____ Date__________

Edison, NJ, 7.00%

Sale, *x*	$20	$30	$40	$50	$60
Sales tax, *T*					

Norman, OK, 7.50%

Sale, *x*	$20	$30	$40	$50	$60
Sales tax, *T*					

What Is Your Answer?

4. **IN YOUR OWN WORDS** How can you use a table to describe a function? Describe an example of a function table in real life.

Amount of Sale	Tax
.10 - .29	.01
.30 - .49	.02
.50 - .69	.03
.70 - .89	.04
.90 - 1.09	.05

"Dear Sir: Yesterday, I bought a piece of 9-cent candy eight times and paid NO tax. Today, I bought eight pieces at once and you charged me $0.04 tax. What's going on?"

Name ______________________________ Date ________

4.3 Practice
For use after Lesson 4.3

Write an equation for the function. Then complete the table.

1. The output is half of the input.

Input, x	0	2	4	6
Output, y				

2. The output is 5 less than the input.

Input, x	5	7	9	11
Output, y				

Write an equation for the function shown by the table.

3.

Input, x	1	2	3	4
Output, y	7	8	9	10

4.

Input, x	0	3	5	9
Output, y	0	6	10	18

Complete the table.

5. For each output, multiply the input by 8, then add 3.

Input, x	1	5		
Output, y	11	43	59	83

6. For each output, divide the input by 4, then subtract 1.

Input, x	4	8		
Output, y	0	1	2	3

7. You take a class to learn sign language. It costs \$8 for each class you attend. Complete the input-output table. Write a function rule in which x is the input and y is the output.

Class, x	1	2	3	4
Cost, y				

Name____ Date__________

4.4 Graphs

For use with Activity 4.4

Essential Question How can you use a graph to describe a function?

1 ACTIVITY: Interpreting a Graph

Work with a partner. Use a graph to test the truth of each statement. If the statement is true, write an equation that shows how to get one measurement from the other measurement.

a. "You can find the horsepower of a race car engine if you know its volume in cubic inches."

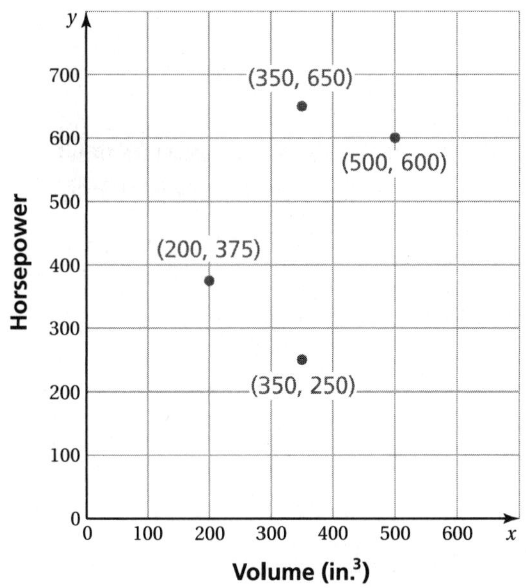

b. "You can find the volume of a race car engine in cubic centimeters if you know its volume in cubic inches."

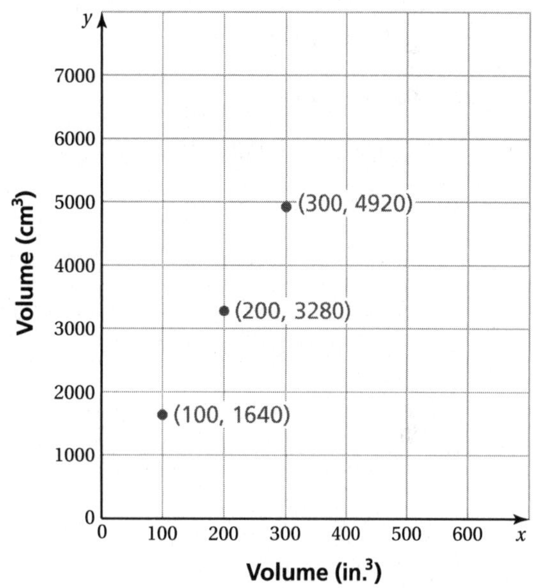

2 ACTIVITY: Interpreting a Graph

Work with a partner. The table shows the average speeds of the winners of the Daytona 500. Graph the data on the next page. Does the graph allow you to predict future winning speeds? Explain why or why not.

Year	2000	2001	2002	2003	2004	2005	2006	2007	2008
Speed (mi/h)	156	162	143	134	156	135	143	149	153

Name ______________________________ Date __________

4.4 Graphs (continued)

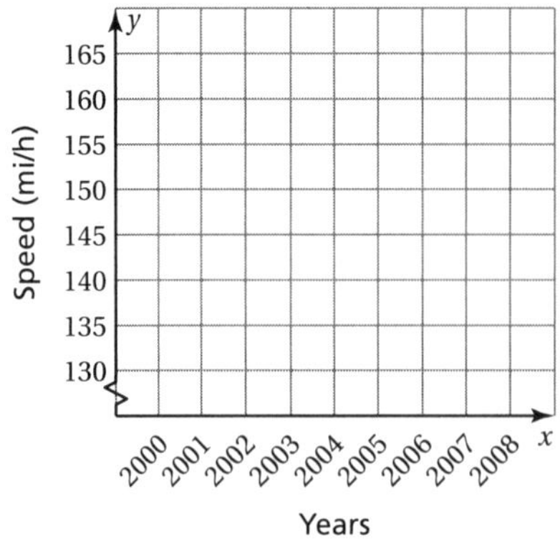

3 ACTIVITY: Conducting an Experiment

Work with a partner.

Collect Materials:

- Metal washer, string (at least 15 in. long), stopwatch

Perform the Experiment:

- Tie one end of the string securely around the washer.
- Hold the string 6 inches from the washer. Swing the washer and measure the time it takes to swing back and forth 10 times.
- Record your result in a table.

String Length (in.)	6	7	8	9	10	11	12
Time (sec)							

- Repeat the experiment when holding the string at lengths of 7 in., 8 in., 9 in., 10 in., 11 in., and 12 in.

Analyze the Results:

- Make a graph of your data.
- Describe the graph.

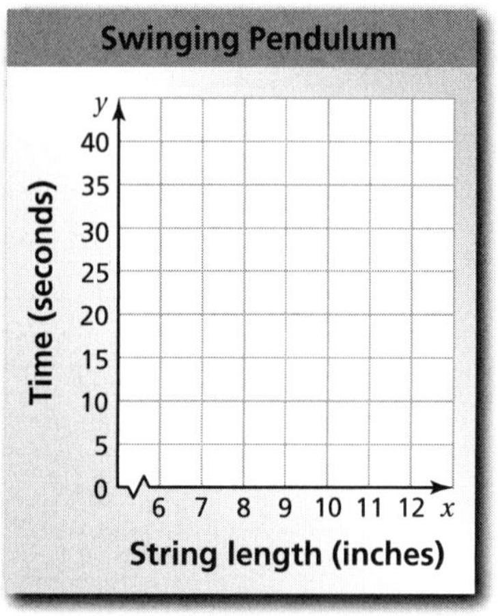

Name____ Date__________

Use Your Results to Predict:

- Use your graph to predict how long it will take a 14-inch pendulum to swing 10 times.

Test Your Prediction:

- Hold the string 14 inches from the washer and repeat the experiment. How close was your prediction?

What Is Your Answer?

4. **IN YOUR OWN WORDS** How can you use a graph to describe a function? Find a graph in a magazine, in a newspaper, or on the Internet that allows you to predict the future.

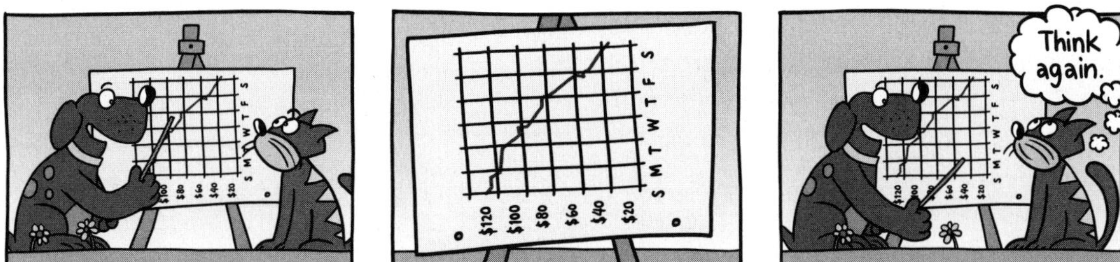

"I graphed our profits."

"And I am happy to say that they are going up every day!"

Name ______________________________ Date __________

4.4 Practice

For use after Lesson 4.4

Graph the data.

1.

Input, *x*	2	4	6	8
Output, *y*	0	2	4	6

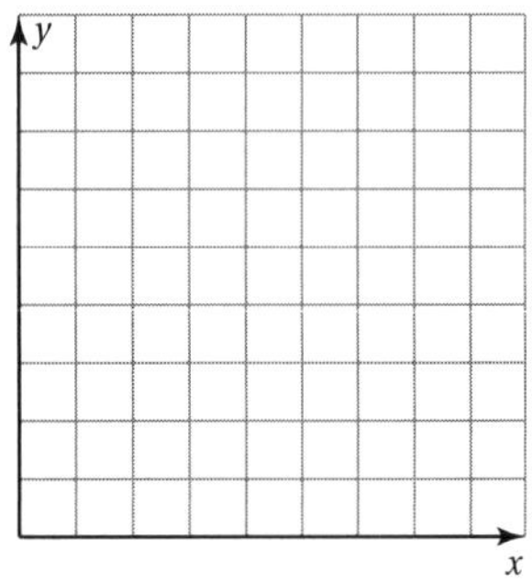

2.

Input, *x*	1	2	3	4
Output, *y*	7	10	13	16

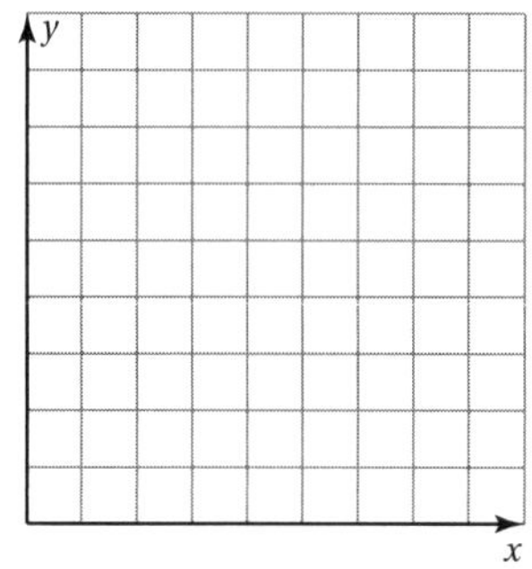

Graph the function.

3. $y = \dfrac{x}{10}$

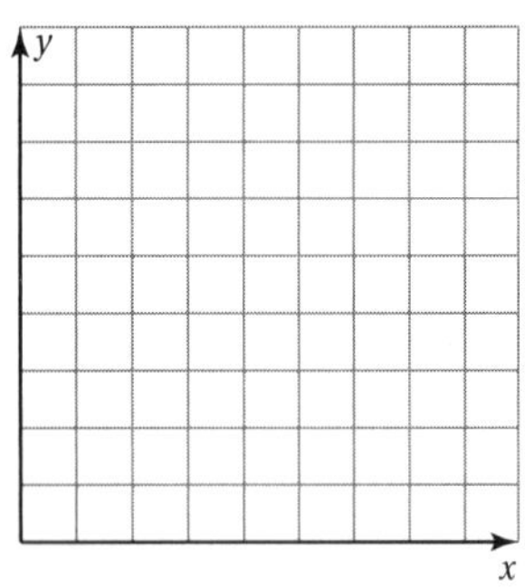

4. $y = \dfrac{1}{2}x + 3$

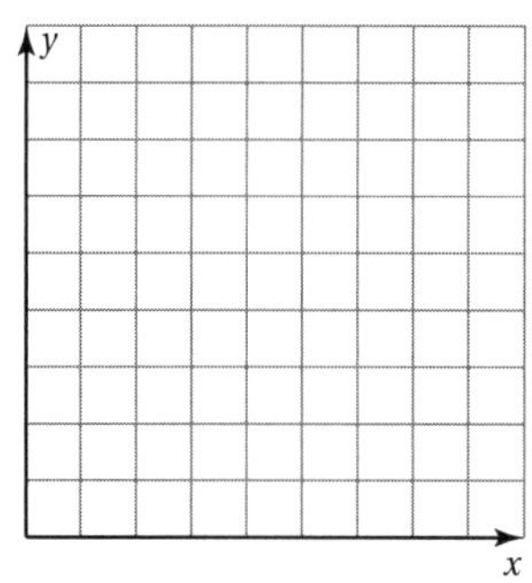

5. While on vacation, you buy key chains for your friends. Each key chain costs \$3. You have a coupon for \$2 off your entire purchase. The function $c = 3k - 2$ relates the number of key chains k you buy and your cost c. Graph the function.

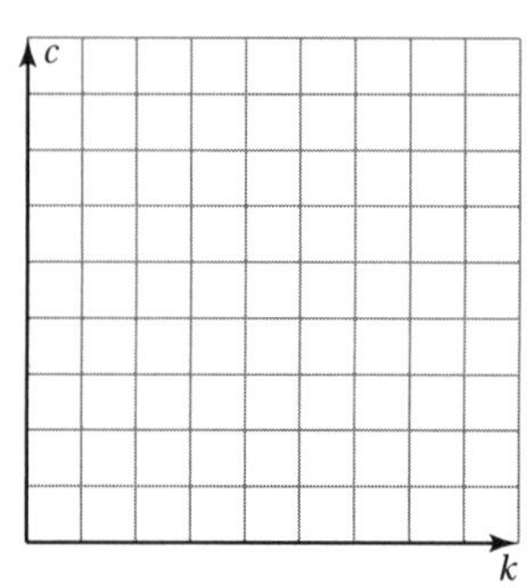

Name____ Date__________

4.5 Analyzing Graphs

For use with Activity 4.5

Essential Question How can you analyze a function from its graph?

1 ACTIVITY: Analyzing Graphs

Work with a partner. Complete the table for the given situation. Then make a graph of the data. Write an equation for the function. Describe the characteristics of the graph.

a. Find the area of a square with side length s.

Side, *s*	1	2	3	4
Area, *A*				

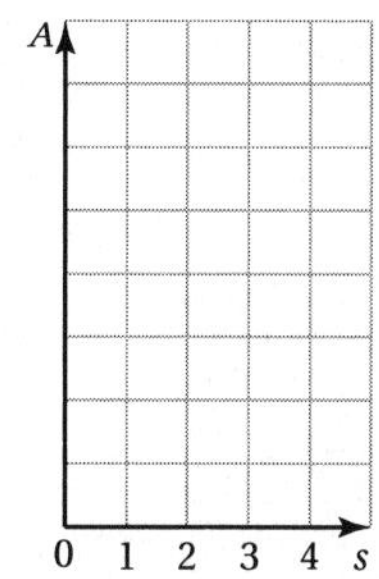

b. Find the amount earned for working h hours at \$3 per hour.

Hour, *h*	1	2	3	4
Amount, *A*				

c. You start with \$20 in a savings account. Find the amount left in the account when you withdraw \$2 each day d.

Day, *d*	1	2	3	4
Amount, *A*				

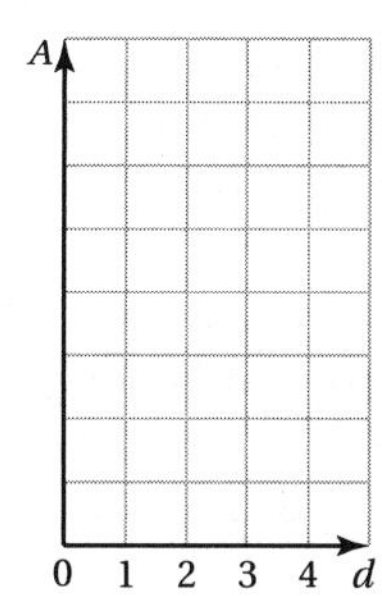

d. You start with \$10 in a savings account. Find the amount in the account when you deposit \$2 each day *d*.

Day, *d*	1	2	3	4
Amount, *A*				

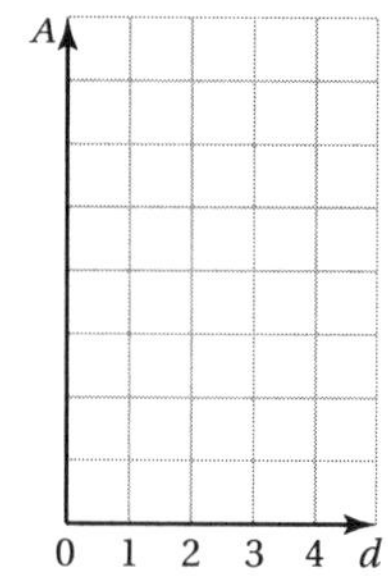

2 ACTIVITY: Conducting an Experiment

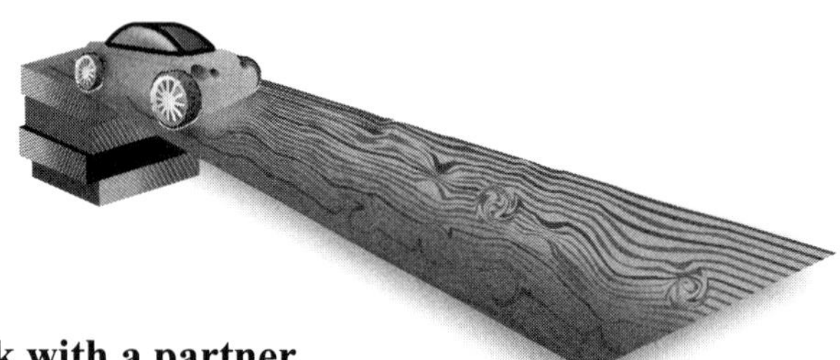

Work with a partner.

Collect Materials:

- A board at least 8 feet long
- Five books of the same thickness
- Toy car
- Stopwatch

Perform the Experiment:

- Place one book underneath one end of the board.
- Put the car at the top of the ramp. Measure the time (in seconds) it takes the car to roll down the ramp.
- Record your result in a table.

Number of Books	1	2	3	4
Time (seconds)				

- Repeat the experiment with two, three, and four books.

4.5 Analyzing Graphs (continued)

Analyze the Results:

- Make a graph of your data.
- Does the graph have the characteristics of any of the graphs in Activity 1? Explain.

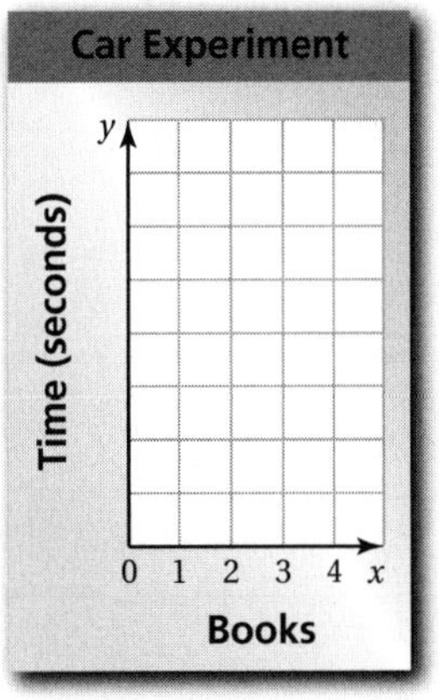

Use Your Results to Predict:

- Use your graph to predict how long it will take the car to roll down the ramp when five books are placed under the board.

Test Your Prediction:

- Repeat the experiment with five books. How close was your prediction?

What Is Your Answer?

3. **IN YOUR OWN WORDS** How can you analyze a function from its graph? Give a real-life example of how a graph can help you make a decision.

Name ______________________________ Date __________

4.5 Practice

For use after Lesson 4.5

Does the graph represent a linear function? Explain.

1.

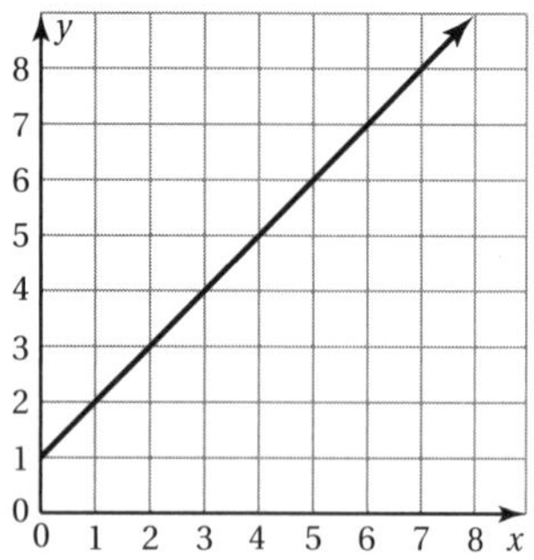

2.

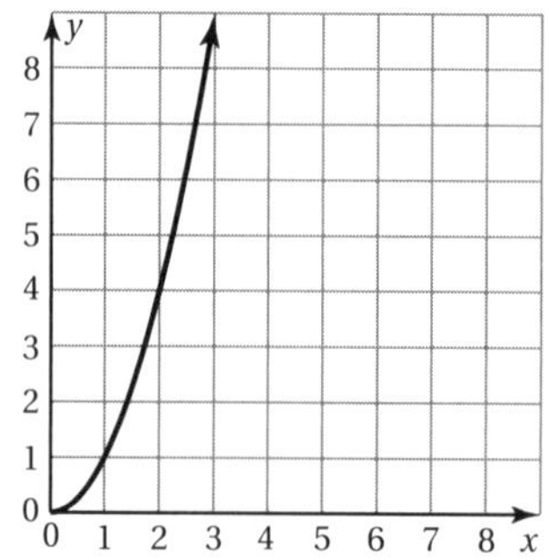

Does the input-output table represent a linear function? Explain.

3.

Input, *x*	0	3	5	9
Output, *y*	0	6	10	18

4.

Input, *x*	1	4	9	16
Output, *y*	1	2	3	4

Graph each linear function. Which graph is steeper? Explain.

5. $y = 3x$ and $y = \frac{1}{3}x$

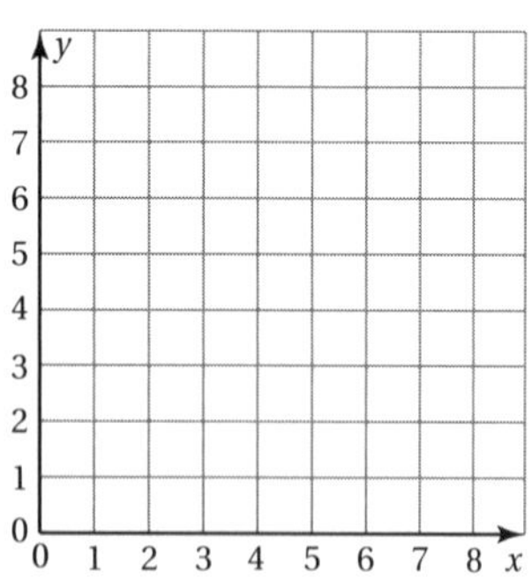

6. $y = \frac{3}{8}x$ and $y = \frac{5}{8}x$

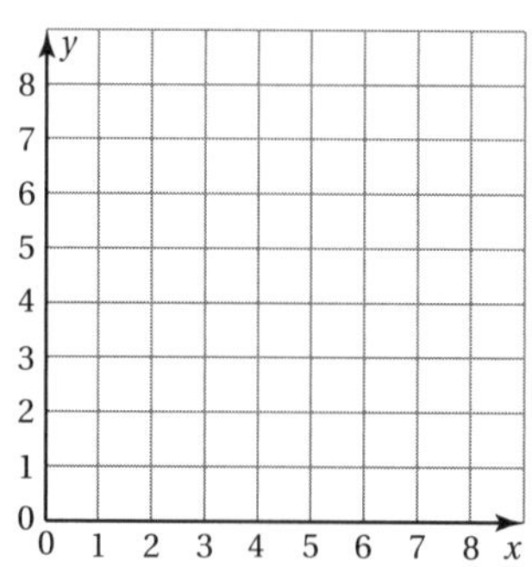

7. The graphs show the number of shirts and pants sales during a week. Which item has more sales in a day? How much more? Explain.

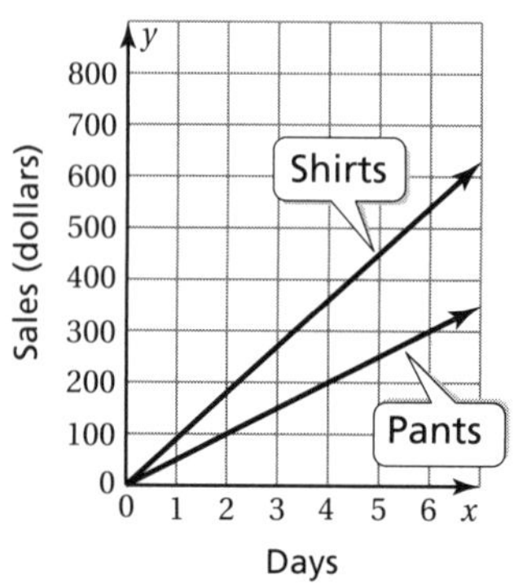

Name____ Date__________

4.6 Arithmetic Sequences
For use with Activity 4.6

Essential Question How are arithmetic sequences used to describe patterns?

1 EXAMPLE: Writing a Story

Write a story that uses the following arithmetic sequence. Draw pictures for your story.

52, 46, 40, …

There are many possible stories. Here is a student's story about temperatures.

Natalie and her family went to a camp in the Blue Ridge Mountains. On Saturday morning she and her family decided to go for a long hike. They started at 8:00 A.M. and by noon had hiked 7 miles.

At noon the temperature was 52 degrees Fahrenheit and a storm started to roll in. Natalie and her family decided to return to their cabin. By 1 P.M. the temperature had dropped to 46 degrees. Natalie said that if the temperature continued to drop at that rate, it would be freezing before they got back to the cabin.

Her father asked how she had come to that conclusion. Here is how she explained it. "We hiked for 4 hours before turning around. If the temperature drops by 6 degrees each hour, the temperature will have dropped to 28 degrees by 4 P.M. Because 32 degrees is freezing, it will be below freezing."

Time	12:00	1:00	2:00	3:00	4:00
Temperature	52°	46°	40°	34°	28°

Name ______________________________ Date ________

2 ACTIVITY: Writing a Story

Work with a partner. Think of a story that uses an arithmetic sequence.

a. Write your story. Then draw pictures for your story.

b. Include a table in your story and show how the table can be used to solve the problem that is described in your story.

You can use any topic and sequence you want. If you can't think of any, here are some possible ideas.

A parachutist falls at the rate of 15 feet per second.

A tree frog jumps 14 inches per jump.

A leaky faucet drips 4 cups of water each hour.

A corn plant grows 4 inches per week.

What Is Your Answer?

3. **IN YOUR OWN WORDS** How are arithmetic sequences used to describe patterns? Give an example from real life.

Name ______________________________ Date __________

4.6 Practice
For use after Lesson 4.6

Write the next three terms of the arithmetic sequence.

1. 24, 13, 2, −9, ...

2. −2.6, −2, −1.4, −0.8, ...

Write a function that describes the arithmetic sequence.

3. 8, 9, 10, 11, ...

4. −5, −10, −15, −20, ...

Write the next three terms of the arithmetic sequence. Then graph the sequence.

5. 16, 14, 12, 10, ...

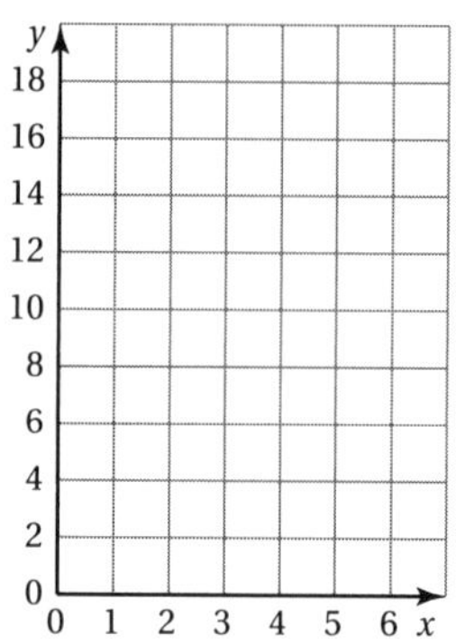

6. 2, 5, 8, 11, ...

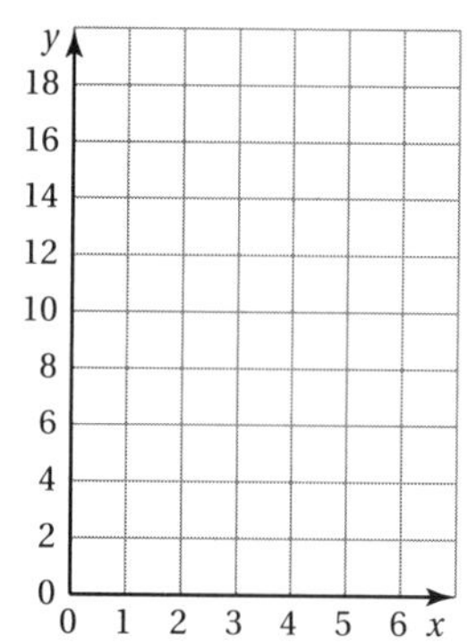

7. The first row in an auditorium has 14 seats. Each row after the first has two more seats than the previous row. Write an arithmetic sequence for the number of seats in each row.

8. The amount you earn each hour for babysitting can be represented by the sequence 8, 16, 24, 32, ... Write a function that describes the arithmetic sequence.

Name ______________________ Date __________

4.7 Geometric Sequences

For use with Activity 4.7

Essential Question How can you find the balance in an account that earns simple interest or compound interest?

1 ACTIVITY: Comparing Simple and Compound Interest

Work with a partner. Interest that is calculated only on principal is *simple interest*. Interest that is calculated on principal *and* previously earned interest is *compound interest*.

You deposit \$1000 in a savings account that earns 6% interest per year. (a) Complete the first table that shows the balance after 10 years with simple interest. (b) Complete the second table that shows the balance after 10 years with interest compounded annually.

$I = Prt = 1000(0.06)(1)$

Simple Interest			
t	Principal	Annual Interest	Balance at End of Year
1	\$1000.00	\$60.00	\$1060.00
2	\$1000.00	\$60.00	\$1120.00
3			
4			
5			
6			
7			
8			
9			
10			

$I = Prt = 1060(0.06)(1)$

Compound Interest			
t	Principal and Interest	Annual Interest	Balance at End of Year
1	\$1000.00	\$60.00	\$1060.00
2	\$1060.00	\$63.60	\$1123.60
3			
4			
5			
6			
7			
8			
9			
10			

Name ______________________________ Date __________

4.7 Geometric Sequences (continued)

c. Which type of interest gives the greater balance?

2 ACTIVITY: Comparing Simple and Compound Interest

Work with a partner.

a. Graph the end-of-year balances for each type of interest in Activity 1.

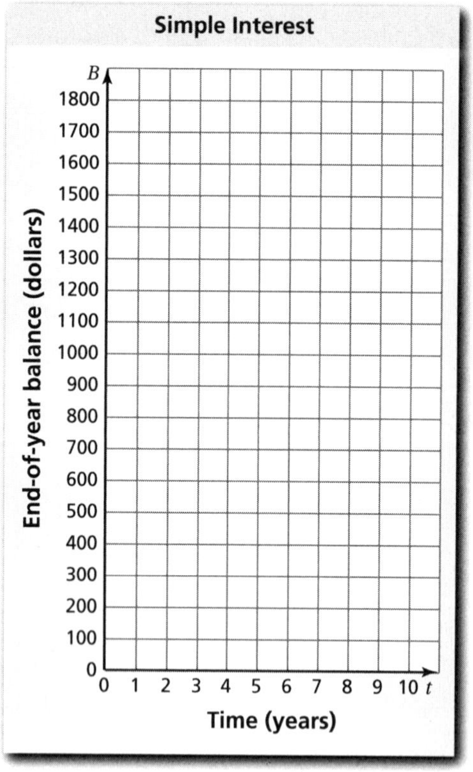

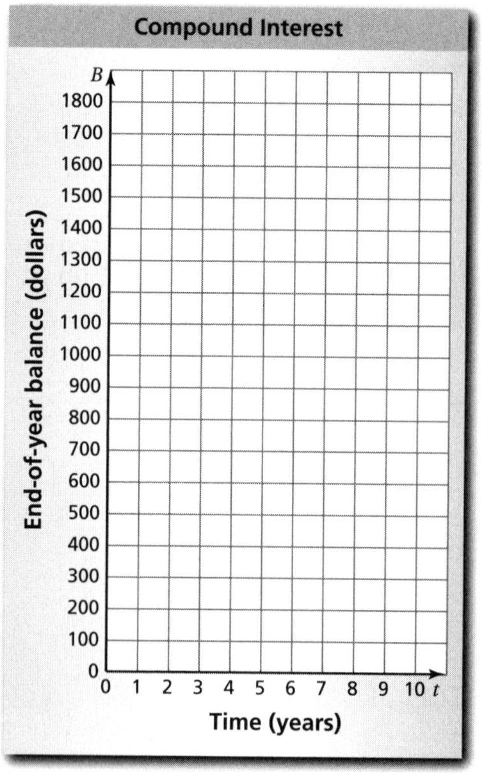

b. Which graph shows an arithmetic sequence? Explain your reasoning.

c. Which graph shows a geometric sequence? Explain your reasoning.

What Is Your Answer?

3. IN YOUR OWN WORDS How can you find the balance in an account that earns simple interest or compound interest?

4. Use what you learned in Activity 2. About how many years will it take for the balance to double with simple interest? with compound interest?

Name ______________________ Date ________

4.7 Practice

For use after Lesson 4.7

Write the next three terms of the geometric sequence.

1. 2, 6, 18, 54, …

2. 64, −16, 4, −1, …

Tell whether the sequence is *geometric, arithmetic,* or *neither*.

3. −53, −33, −13, 7, …

4. 2, −4, 6, −8, …

5. 128, −64, 32, −16, …

Write the next three terms of the geometric sequence. Then graph the sequence.

6. 81, 27, 9, 3, …

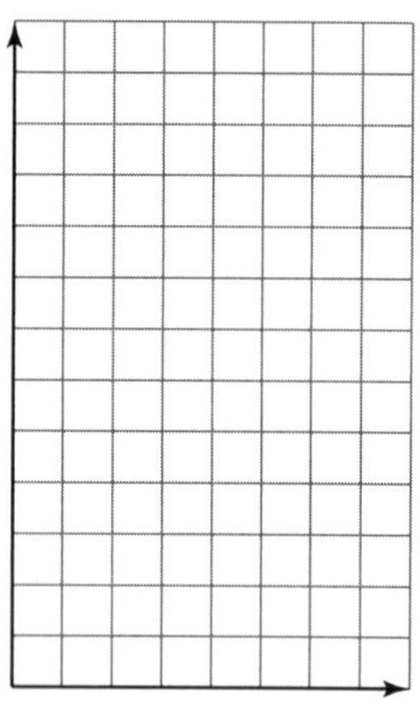

7. $\frac{1}{64}, \frac{1}{16}, \frac{1}{4}, 1, \ldots$

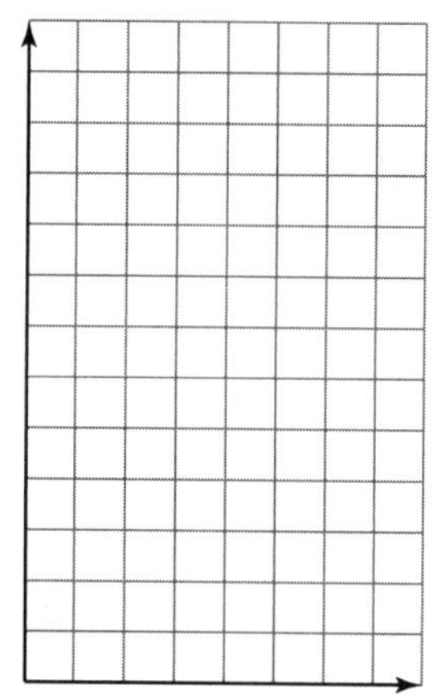

8. The first two terms of a sequence are −1 and −4.

a. Suppose the sequence is geometric. Write the next three terms.

b. Suppose the sequence is arithmetic. Write the next three terms.

Name____ Date__________

Chapter 5 Fair Game Review

Simplify.

1. $\frac{3}{18}$

2. $\frac{4}{6}$

3. $\frac{12}{60}$

4. $\frac{14}{28}$

5. $\frac{16}{36}$

6. $\frac{40}{50}$

Are the fractions equivalent?

7. $\frac{3}{8} \stackrel{?}{=} \frac{6}{11}$

8. $\frac{4}{10} \stackrel{?}{=} \frac{16}{40}$

9. $\frac{22}{32} \stackrel{?}{=} \frac{11}{16}$

10. $\frac{63}{72} \stackrel{?}{=} \frac{7}{9}$

11. You see 58 birds while on a bird watching tour. Of those birds, you see 12 hawks. Write and simplify the fraction of hawks you see.

Name ______________________________ Date __________

Chapter 5 Fair Game Review (continued)

Solve the equation. Check your solution.

12. $\frac{d}{12} = -4$

13. $-7 = \frac{x}{-3}$

14. $\frac{1}{8}n = 5$

15. $6a = -54$

16. $10 = -2k$

17. $2.7 = -0.9y$

18. $-23.4 = -1.3w$

19. $\frac{1}{15}z = 6$

20. You and three friends spend $35 on tickets at the movies. Write and solve an equation to find the price p of one ticket.

Name ______________________________ Date __________

5.1 Ratios and Rates

For use with Activity 5.1

Essential Question How do rates help you describe real-life problems?

1 ACTIVITY: Finding Reasonable Rates

Work with a partner.

a. Match each description with a verbal rate.

b. Match each verbal rate with a numerical rate.

c. Give a reasonable numerical rate for each description. Then give an unreasonable rate.

Description	*Verbal Rate*	*Numerical Rate*
Your pay rate for washing cars	inches per month	$\frac{\square \text{ m}}{\text{sec}}$; $\frac{\square \text{ m}}{\text{sec}}$
The average rainfall rate in a rain forest	pounds per acre	$\frac{\square \text{ people}}{\text{yr}}$; $\frac{\square \text{ people}}{\text{yr}}$
Your average driving rate along an interstate	meters per second	$\frac{\square \text{ lb}}{\text{acre}}$; $\frac{\square \text{ lb}}{\text{acre}}$
The growth rate for the length of a baby alligator	people per year	$\frac{\square \text{ mi}}{\text{h}}$; $\frac{\square \text{ mi}}{\text{h}}$
Your running rate in a 100-meter dash	dollars per hour	$\frac{\square \text{ in.}}{\text{yr}}$; $\frac{\square \text{ in.}}{\text{yr}}$
The population growth rate of a large city	dollars per year	$\frac{\square \text{ in.}}{\text{mo}}$; $\frac{\square \text{ in.}}{\text{mo}}$
The average pay rate for a professional athlete	miles per hour	$\frac{\$\square}{\text{h}}$; $\frac{\$\square}{\text{h}}$
The fertilization rate for an apple orchard	inches per year	$\frac{\$\square}{\text{yr}}$; $\frac{\$\square}{\text{yr}}$

5.1 Ratios and Rates (continued)

2 ACTIVITY: Unit Analysis

Work with a partner. Some real-life problems involve the product of an amount and a rate. Find each product. List the units.

a. $6\text{ h} \times \dfrac{\$12}{\text{h}}$

b. $6\text{ mo} \times \dfrac{\$700}{\text{mo}}$

c. $10\text{ gal} \times \dfrac{22\text{ mi}}{\text{gal}}$

d. $9\text{ lb} \times \dfrac{\$3}{\text{lb}}$

e. $13\text{ min} \times \dfrac{60\text{ sec}}{\text{min}}$

3 ACTIVITY: Writing a Story

Work with a partner.

- **Think of a story that compares two different rates.**
- **Write the story.**
- **Draw pictures for the story.**

What Is Your Answer?

4. **RESEARCH** Use newspapers, the Internet, or magazines to find examples of salaries. Try to find examples of each of the following ways to write salaries.

   a. dollars per hour b. dollars per month c. dollars per year

5. **IN YOUR OWN WORDS** How do rates help you describe real-life problems? Give two examples.

6. To estimate the annual salary for a given hourly pay rate, multiply by 2 and insert "000" at the end. **Sample:** \$10 per hour is about \$20,000 per year.

   a. Explain why this works. Assume the person is working 40 hours a week.

   b. Estimate the annual salary for an hourly pay rate of \$8 per hour.

   c. You earn \$1 million per month. What is your annual salary?

   d. Why is the cartoon funny?

"We had someone apply for the job. He says he would like \$1 million a month, but will settle for \$8 an hour."

Name ______________________________ Date __________

5.1 Practice

For use after Lesson 5.1

Write the ratio as a fraction in simplest form.

1. 8 to 14

2. 36 even : 12 odd

3. 42 vanilla to 48 chocolate

Find the unit rate.

4. \$2.50 for 5 ounces

5. 15 degrees in 2 hours

6. 183 miles in 3 hours

Use the table to find the rate.

7.

Boxes	0	1	2	3
Pounds	0	30	60	90

8.

Notebooks	0	5	10	15
Dollars	0	9.45	18.90	28.35

9. A laser printer prints 360 pages in 30 minutes. What is the printing rate in pages per minute?

10. A clothing store sells four shirts for \$60.00. The next week, the store runs a special that is buy three shirts for \$19.50 each, get the fourth free. Which is the better buy?

11. You create 15 centerpieces for a party in 5 hours.

a. What is the unit rate?

b. How long will it take you to make 42 centerpieces?

Name_______________________________________ Date__________

5.2 Slope

For use with Activity 5.2

Essential Question How can you compare two rates graphically?

1 ACTIVITY: Comparing Unit Rates

Work with a partner. The table shows the maximum speeds of several animals.

a. Find the missing speeds. Round your answers to the nearest tenth.

b. Which animal is fastest? Which animal is slowest?

c. Explain how you convert between the two units of speed.

Animal	Speed (miles per hour)	Speed (feet per second)
Antelope	61.0	
Black Mamba Snake		29.3
Cheetah		102.6
Chicken		13.2
Coyote	43.0	
Domestic Pig		16.0
Elephant		36.6
Elk		66.0
Giant Tortoise	0.2	
Giraffe	32.0	
Gray Fox		61.6
Greyhound	39.4	
Grizzly Bear		44.0
Human		41.0
Hyena	40.0	
Jackal	35.0	
Lion		73.3
Peregrine Falcon	200.0	
Quarter Horse	47.5	
Spider		1.76
Squirrel	12.0	
Thomson's Gazelle	50.0	
Three-Toed Sloth		0.2
Tuna	47.0	

Name ____ Date __________

2 ACTIVITY: Comparing Two Rates Graphically

Work with a partner. A cheetah and a Thomson's gazelle are running at constant speeds.

a. Find the missing distances.

	Cheetah	Gazelle
Time (seconds)	**Distance (feet)**	**Distance (feet)**
0	0	0
1	102.6	
2		
3		
4		
5		
6		
7		

b. Use the table to complete the line graph for each animal.

c. Which graph is steeper? The speed of which animal is greater?

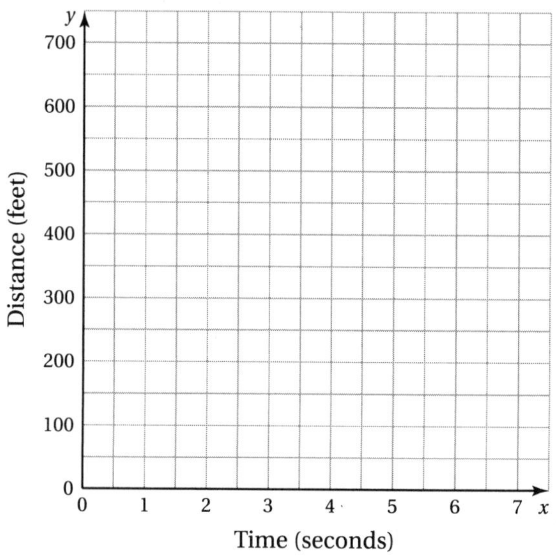

Name______________________________ Date__________

5.2 **Slope (continued)**

What Is Your Answer?

3. **IN YOUR OWN WORDS** How can you compare two rates graphically? Explain your reasoning. Give some examples with your answer.

4. Choose 10 animals from Activity 1.

   a. Make a table for each animal similar to the table in Activity 2.

   b. Sketch a graph of the distances for each animal.

   c. Compare the steepness of the 10 graphs. What can you conclude?

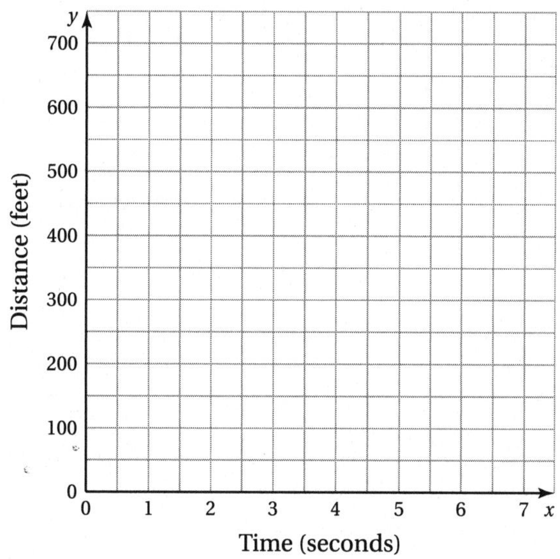

Name ________________________________ Date __________

5.2 Practice

For use after Lesson 5.2

Find the slope of the line.

1. 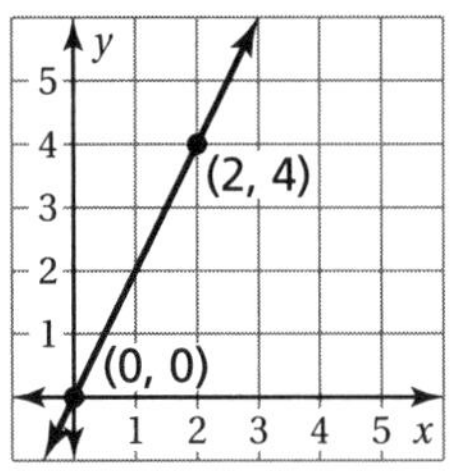

2. 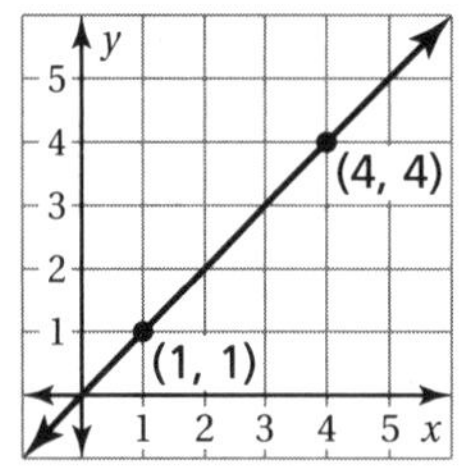

3. 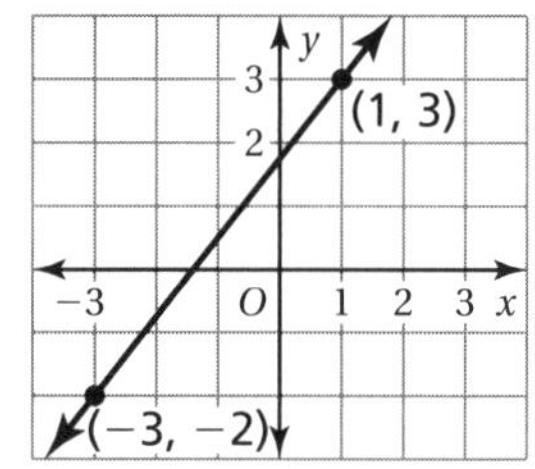

Graph the data. Then find the slope of the line through the points.

4.

Minutes, *x*	0	1	3	5
Pages, *y*	0	1.5	4.5	7.5

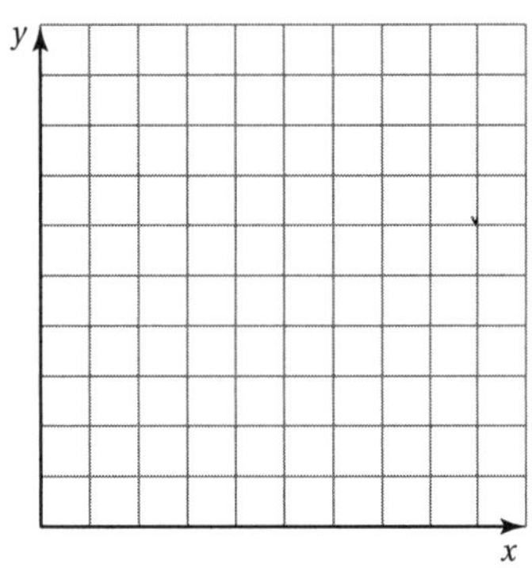

5.

Miles, *x*	0	1	2	3
Calories, *y*	0	135	270	405

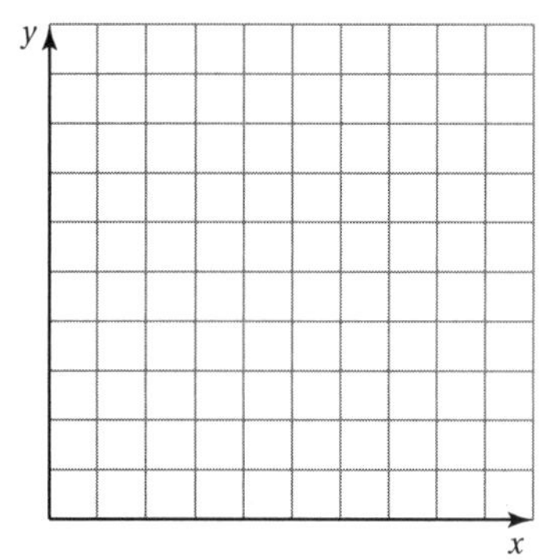

6. By law, the maximum slope of a wheelchair ramp is $\frac{1}{12}$.

a. A ramp is designed that is 4 feet high and has a horizontal length of 50 feet. Does this ramp meet the law? Explain.

b. What could be adjusted on an unacceptable ramp so that it meets the law?

Name____ Date__________

5.3 Proportions

For use with Activity 5.3

Essential Question How can proportions help you decide when things are "fair"?

1 ACTIVITY: Determining Proportions

Work with a partner. Tell whether the two ratios are equivalent. If they are not equivalent, change the second day to make the ratios equivalent. Explain your reasoning.

a. On the first day, you pay \$5 for 2 boxes of popcorn. The next day, you pay \$7.50 for 3 boxes.

First Day

$$\frac{\$5.00}{\$7.50} \stackrel{?}{=} \frac{2 \text{ boxes}}{3 \text{ boxes}}$$

Next Day

b. On the first day, it takes you 3 hours to drive 135 miles. The next day, it takes you 5 hours to drive 200 miles.

First Day

$$\frac{3 \text{ h}}{5 \text{ h}} \stackrel{?}{=} \frac{135 \text{ mi}}{200 \text{ mi}}$$

Next Day

c. On the first day, you walk 4 miles and burn 300 calories. The next day, you walk 3 miles and burn 225 calories.

First Day

$$\frac{4 \text{ mi}}{3 \text{ mi}} \stackrel{?}{=} \frac{300 \text{ cal}}{225 \text{ cal}}$$

Next Day

d. On the first day, you download 5 songs and pay \$2.25. The next day, you download 4 songs and pay \$2.00.

First Day

$$\frac{5 \text{ songs}}{4 \text{ songs}} \stackrel{?}{=} \frac{\$2.25}{\$2.00}$$

Next Day

2 ACTIVITY: Checking a Proportion

Work with a partner.

a. It is said that "one year in a dog's life is equivalent to seven years in a human's life." Explain why Newton thinks he has a score of 105 points. Did he solve the proportion correctly?

$$\frac{1 \text{ year}}{7 \text{ years}} \stackrel{?}{=} \frac{15 \text{ points}}{105 \text{ points}}$$

"I got 15 on my online test. That's 105 in dog points! Isn't that an A+?"

b. If Newton thinks his score is 98 points, how many points does he actually have? Explain your reasoning.

3 ACTIVITY: Determining Fairness

Work with a partner. Write a ratio for each sentence. If they are equal, then the answer is "It is fair." If they are not equal, then the answer is "It is not fair." Explain your reasoning.

a. You pay \$184 for 2 tickets to a concert. & I pay \$266 for 3 tickets to the same concert. → **Is this fair?**

b.

You get 75 points for answering 15 questions correctly.	&	I get 70 points for answering 14 questions correctly.	➡ **Is this fair?**

c.

You trade 24 football cards for 15 baseball cards.	&	I trade 20 football cards for 32 baseball cards.	➡ **Is this fair?**

What Is Your Answer?

4. Find a recipe for something you like to eat. Then show how two of the ingredient amounts are proportional when you double or triple the recipe.

5. **IN YOUR OWN WORDS** How can proportions help you decide when things are “fair?” Give an example.

Name ______________________________ Date __________

5.3 Practice

For use after Lesson 5.3

Tell whether the ratios form a proportion.

1. $\frac{1}{5}, \frac{5}{15}$

2. $\frac{2}{3}, \frac{12}{18}$

3. $\frac{15}{2}, \frac{4}{30}$

4. $\frac{56}{21}, \frac{8}{3}$

5. $\frac{5}{8}, \frac{62.5}{100}$

6. $\frac{17}{20}, \frac{90.1}{106}$

7. $\frac{3.2}{4}, \frac{16}{24}$

8. $\frac{34}{50}, \frac{6.8}{10}$

Tell whether the two rates form a proportion.

9. 28 points in 3 games; 112 points in 12 games

10. 32 notes in 4 measures; 12 notes in 2 measures

11. You can type 105 words in two minutes. Your friend can type 210 words in four minutes. Are these rates proportional? Explain.

12. You make punch for a party. The ratio of ginger ale to fruit juice is 8 cups to 3 cups. You decide to add 4 more cups of ginger ale. How many more cups of fruit juice do you need to add to keep the correct ratio? Explain.

Name______________________________________ Date__________

5.4 Writing Proportions

For use with Activity 5.4

Essential Question How can you write a proportion that solves a problem in real life?

1 ACTIVITY: Writing Proportions

Work with a partner. A rough rule for finding the correct bat length is "The bat length should be half of the batter's height." So, a 62-inch tall batter uses a bat that is 31 inches long. Write a proportion to find the bat length for each given batter height.

a. 58 inches **b.** 60 inches **c.** 64 inches

2 ACTIVITY: Bat Lengths

Work with a partner. Here is a more accurate table for determining the bat length for a batter. Find all of the batter heights for which the rough rule in Activity 1 is exact.

Weight of Batter (pounds)	Height of Batter (inches)							
	45–48	**49–52**	**53–56**	**57–60**	**61–64**	**65–68**	**69–72**	**Over 72**
Under 61	28	29	29					
61–70	28	29	30	30				
71–80	28	29	30	30	31			
81–90	29	29	30	30	31	32		
91–100	29	30	30	31	31	32		
101–110	29	30	30	31	31	32		
111–120	29	30	30	31	31	32		
121–130	29	30	30	31	32	33	33	
131–140	30	30	31	31	32	33	33	
141–150	30	30	31	31	32	33	33	
151–160	30	31	31	32	32	33	33	33
161–170		31	31	32	32	33	33	34
171–180				32	33	33	34	34
Over 180					33	33	34	34

Name ______________________________ Date __________

5.4 Writing Proportions (continued)

3 ACTIVITY: Writing Proportions

Work with a partner. The batting average of a baseball player is the number of "hits" divided by the number of "at bats."

$$\text{Batting Average} = \frac{\text{Hits } (H)}{\text{At Bats } (A)}$$

A player whose batting average is 0.250 is said to be "batting 250."

$$\frac{20 \text{ hits}}{80 \text{ at bats}} = 0.250 = \frac{250 \text{ hits}}{1000 \text{ at bats}}$$

Actual hits

Actual at bats

Batting average

Batting 250 out of 1000

Write a proportion to find how many hits *H* a player needs to achieve the given batting average. Then solve the proportion.

a. 50 times at bat;
batting average is 0.200.

b. 84 times at bat;
batting average is 0.250.

c. 80 times at bat;
batting average is 0.350.

d. 1 time at bat;
batting average is 1.000.

What Is Your Answer?

4. IN YOUR OWN WORDS How can you write a proportion that solves a problem in real life?

5. Two players have the same batting average.

	At Bats	Hits	Batting Average
Player 1	132	45	
Player 2	132	45	

Player 1 gets four hits in the next five at bats. Player 2 gets three hits in the next three at bats.

a. Who has the higher batting average?

b. Does this seem fair? Explain your reasoning.

Name ______________________________ Date __________

5.4 Practice

For use after Lesson 5.4

Write a proportion to find how many points a student needs to score on the test to get the given score.

1. Test worth 50 points; test score of 84%

2. Test worth 75 points; test score of 96%

Use the table to write a proportion.

3.

	Trip 1	Trip 2
Miles	104	78
Gallons	4	g

4.

	Tree 1	Tree 2
Inches	15	x
Years	4	3

Solve the proportion.

5. $\frac{1}{3} = \frac{x}{12}$

6. $\frac{5}{9} = \frac{25}{y}$

7. $\frac{26}{z} = \frac{13}{22}$

8. $\frac{b}{30} = \frac{2.6}{1.5}$

9. A local Humane Society houses 300 animals. The ratio of cats to all animals is 7 : 15.

a. Write a proportion that gives the number of cats c.

b. How many cats are in the Humane Society?

10. Your school buys 30 graphing calculators for $1822.50. Write and solve a proportion that gives the cost c of buying 120 calculators.

Name____ Date__________

5.5 Solving Proportions

For use with Activity 5.5

Essential Question How can you use ratio tables and cross products to solve proportions in science?

1 ACTIVITY: Solving a Proportion in Science

SCIENCE Scientists use *ratio tables* to determine the amount of a compound (like salt) that is dissolved in a solution. Work with a partner to show how scientists use cross products to determine the unknown quantity in a ratio.

a. Sample: Salt Water

Salt Water	1 L	3 L
Salt	250 g	x g

I liter 3 liter

There are _____ grams of salt in the 3-liter solution.

b. White Glue Solution

Water	$\frac{1}{2}$ cup	1 cup
White Glue	$\frac{1}{2}$ cup	x cups

c. Borax Solution

Borax	1 tsp	2 tsp
Water	1 cup	x cups

Name ____ Date __________

d. Slime (see recipe)

Borax Solution	$\frac{1}{2}$ cup	1 cup
White Glue Solution	y cups	x cups

2 ACTIVITY: The Game of Criss Cross

Preparation:

- Cut index cards to make 48 playing cards.
- Write each number on a card.
  1, 1, 1, 2, 2, 2, 3, 3, 3, 4, 4, 4, 5, 5, 5, 6, 6, 6, 7, 7, 7, 8, 8, 8, 9, 9, 9, 10, 10, 10, 12, 12, 12, 13, 13, 13, 14, 14, 14, 15, 15, 15, 16, 16, 16, 18, 20, 25
- Make a copy of the game board.

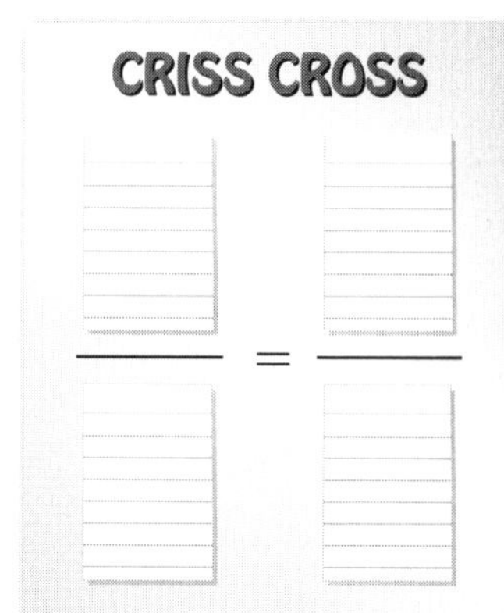

To Play:

- Play with a partner.
- Deal 8 cards to each player.
- Begin by drawing a card from the remaining cards. Use four of your cards to try and form a proportion.
- Lay the four cards on the game board. If you form a proportion, say "Criss Cross" and you earn 4 points. Place the four cards in a discard pile. Now it is your partner's turn.
- If you cannot form a proportion, then it is your partner's turn.
- When the original pile of cards is empty, shuffle the cards in the discard pile and start again.
- The first player to reach 20 points wins.

Name____ Date__________

5.5 Solving Proportions (continued)

What Is Your Answer?

3. IN YOUR OWN WORDS How can you use ratio tables and cross products to solve proportions in science? Give an example.

4. PUZZLE Use each number once to form three proportions.

1	2	10	4	12	20
15	5	16	6	8	3

Name ____ Date __________

5.5 Practice

For use after Lesson 5.5

Solve the proportion using multiplication.

1. $\frac{a}{40} = \frac{3}{10}$ **2.** $\frac{6}{11} = \frac{c}{77}$ **3.** $\frac{b}{65} = \frac{7}{13}$

Solve the proportion using the Cross Products Property.

4. $\frac{k}{6} = \frac{8}{16}$ **5.** $\frac{5.4}{7} = \frac{27}{h}$ **6.** $\frac{8}{11} = \frac{4}{y + 2}$

Write and solve a proportion to complete the statement.

7. 42 in. $\approx$ ______ cm **8.** 12.6 kg $\approx$ ______ lb **9.** 3 oz $\approx$ ______ g

10. A cell phone company charges $5 for 250 text messages. How much does the company charge for 300 text messages?

11. There are 84 players on a football team. The ratio of offensive players to defensive players is 4 to 3. How many offensive players are on the team?

Name ______________________________ Date __________

Chapter 6 Fair Game Review

Find the perimeter or circumference.

1.

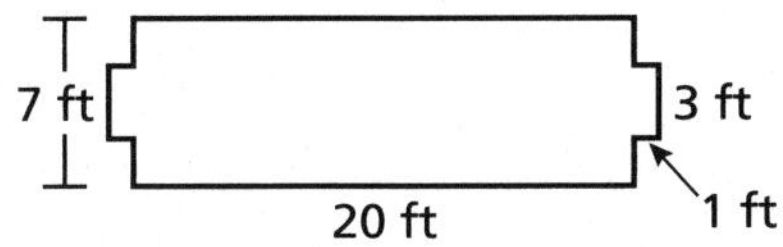

2.

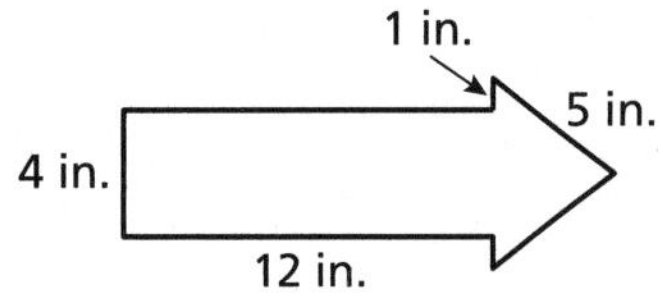

3.

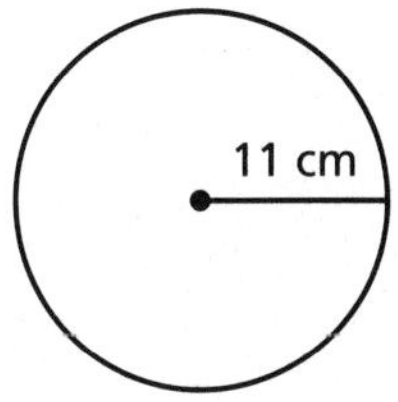

4.

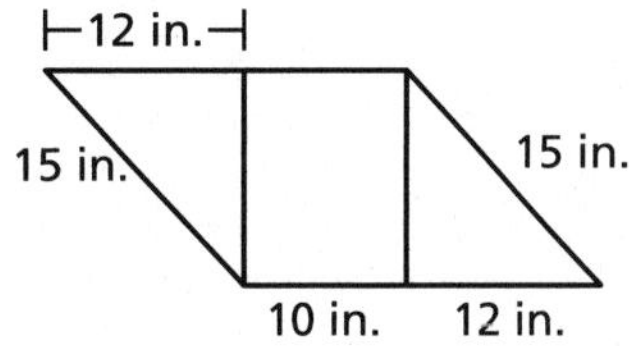

5.

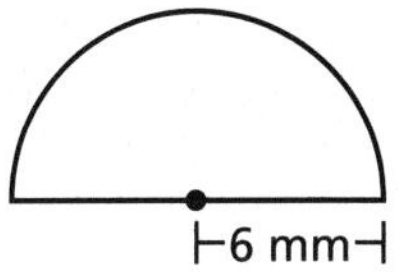

6.

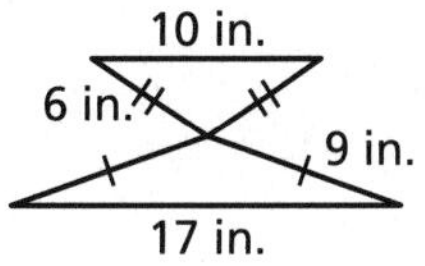

7. A restaurant sets up tables for a party in a U-shape. What is the perimeter of the tables?

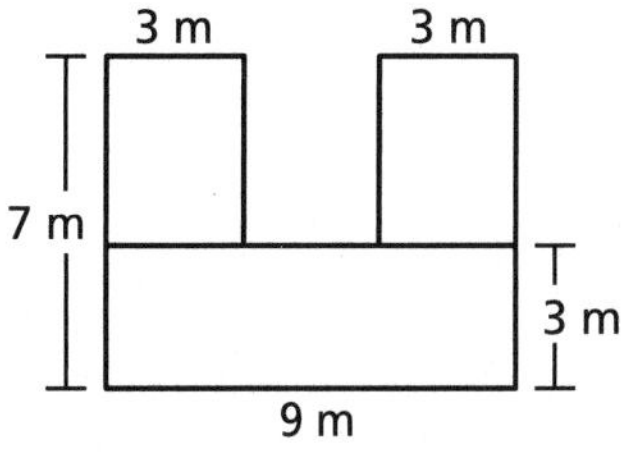

Name ______________________________ Date __________

Chapter 6 Fair Game Review (continued)

Solve the proportion.

8. $\frac{x}{20} = \frac{4}{5}$

9. $\frac{6}{x} = \frac{9}{12}$

10. $\frac{4}{9} = \frac{5}{x}$

11. $\frac{2x}{21} = \frac{2}{7}$

12. $\frac{18}{5x} = \frac{3}{5}$

13. $\frac{9}{10} = \frac{108}{10x}$

14. A flower shop sells a dozen roses for $25. How much does it cost to buy 18 roses?

Name____ Date__________

6.1 Quadrilaterals

For use with Activity 6.1

Essential Question What are the properties of the angles of special types of quadrilaterals?

1 ACTIVITY: Classifying Quadrilaterals

Work with a partner. Label each quadrilateral as a square, rectangle, parallelogram, trapezoid, rhombus, kite, or simply as a quadrilateral. (Use the name that is the most specific.) Cut the shapes to form 12 puzzle pieces.* Arrange the pieces so that they form a square.

*Cut-outs are available in the back of the Record and Practice Journal.

2 ACTIVITY: Properties of Quadrilaterals

Work with a partner. Use a protractor to measure the angles of each puzzle piece in Activity 1. Label each angle. Talk about any properties that you observe. Then answer each of the following.

a. Describe a property of the measures of the angles of a square.

b. Describe a property of the measures of the angles of a rectangle.

c. Describe a property of the measures of opposite angles of a parallelogram.

d. Describe a property of the measures of adjacent angles of a parallelogram.

e. The Greek word *isos* means equal. The Greek word *skelos* means leg. Write a definition for an isosceles trapezoid. Then, describe a property of the measures of the angles of an isosceles trapezoid.

f. Describe a property of the measures of the angles of a kite.

g. Write a definition for a right trapezoid. Describe a property of the measures of the two nonright angles of a right trapezoid.

What Is Your Answer?

3. IN YOUR OWN WORDS What are the properties of the angles of special types of quadrilaterals? Organize your work in a table.

4. Describe how knowing properties of the angles of special types of quadrilaterals is important in a construction career.

Name ______________________ Date ________

6.1 Practice
For use after Lesson 6.1

Identify the quadrilateral.

1. 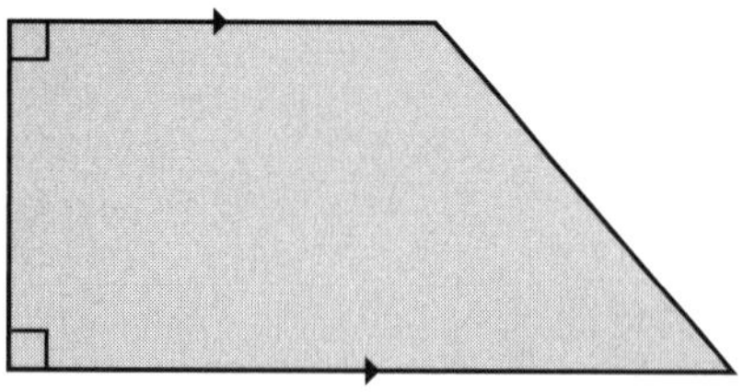

2. 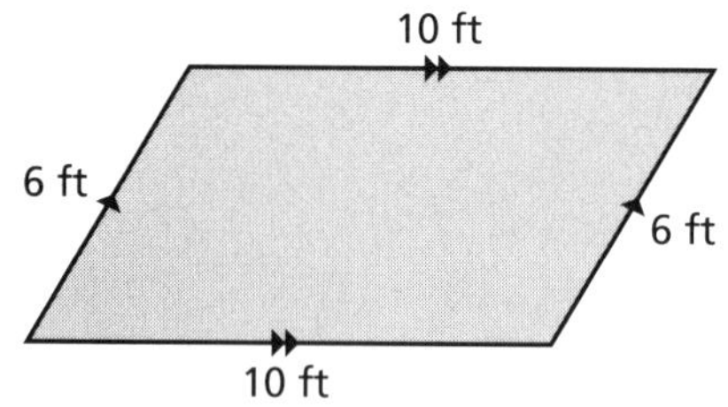

Find the missing angle measure(s).

3. 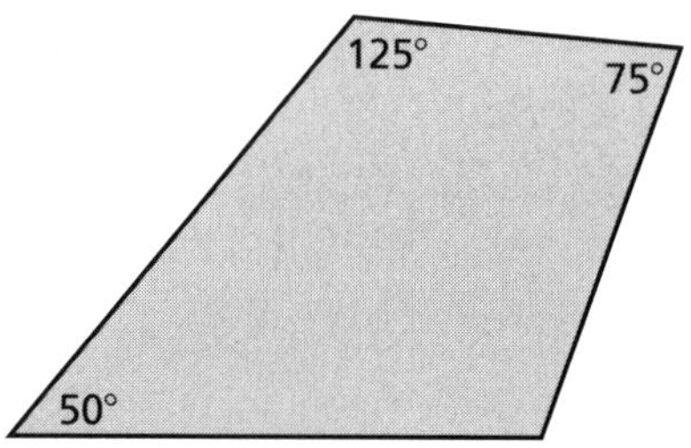

4. 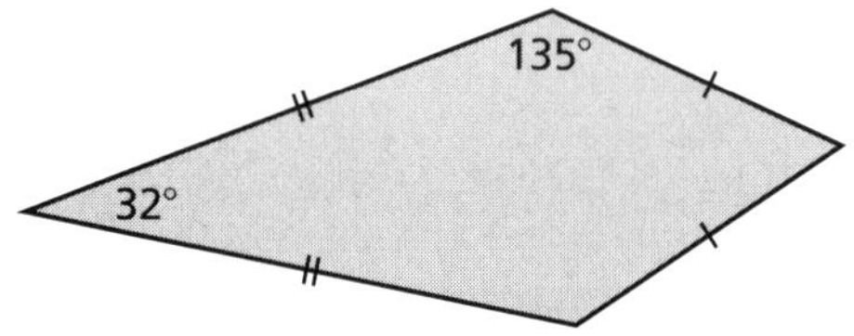

5. 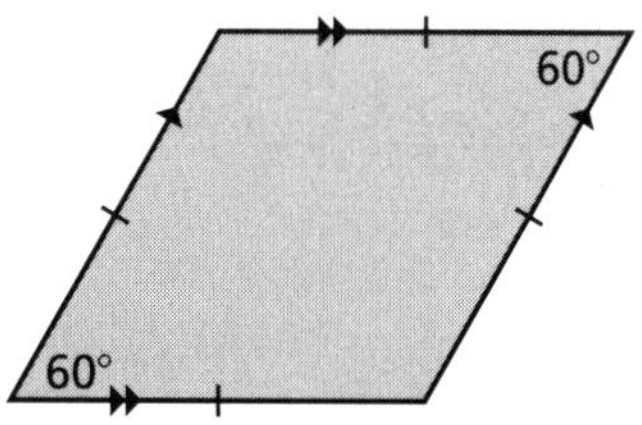

6. 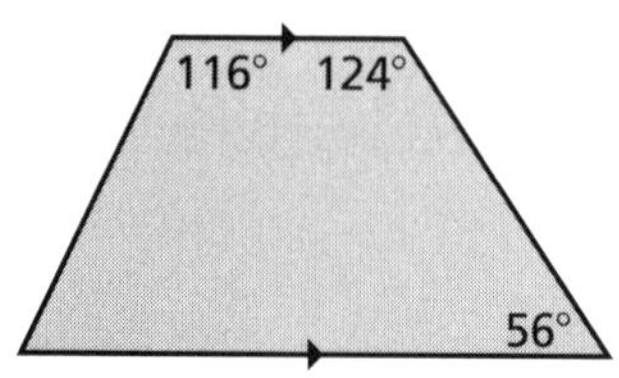

7. A surveyor is doing an analysis on four roads and their intersections.

   a. Identify the shape of the four roads.

   b. Find the missing angle measures.

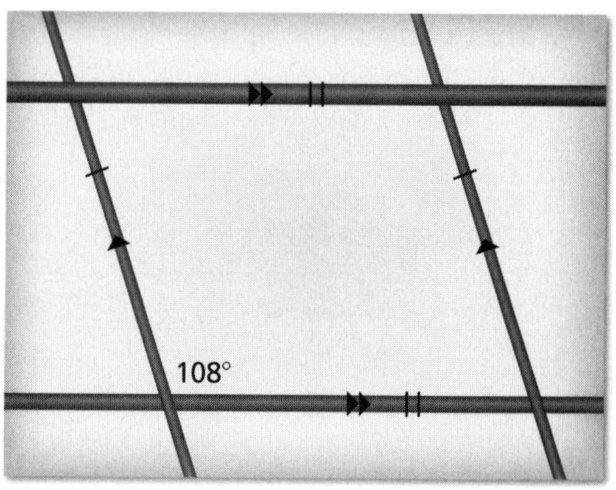

Name________________________________ Date__________

6.2 Identifying Similar Figures

For use with Activity 6.2

Essential Question How can you use proportions to help make decisions in art, design, and magazine layouts?

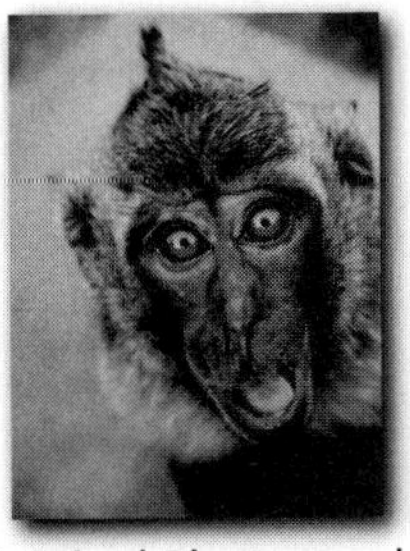

Original Photograph

In a computer art program, when you click and drag on a side of a photograph, you distort it.

But when you click and drag on a corner of the photograph, it remains proportional to the original.

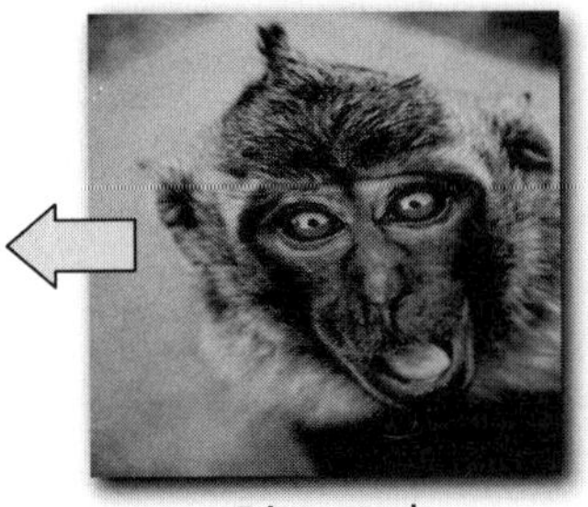

Distorted

Distorted

Proportional

1 ACTIVITY: Reducing Photographs

Work with a partner. You are trying to reduce the photograph to the indicated size for a nature magazine. Can you reduce the photograph to the indicated size without distorting or cropping? Explain your reasoning.

a.

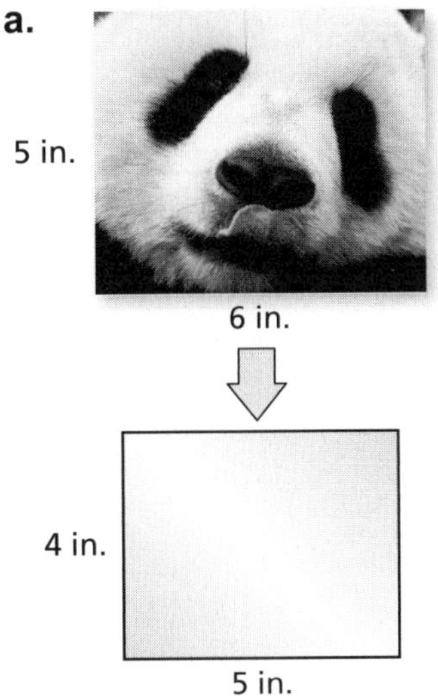

b.

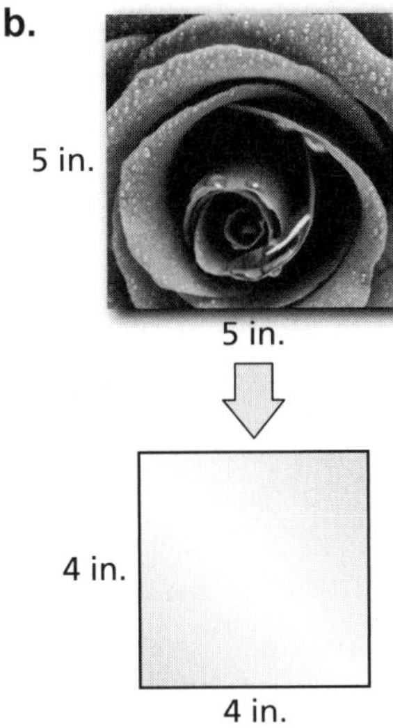

c.

Name ____ Date __________

6.2 Identifying Similar Figures (continued)

2 ACTIVITY: Proportional Designs

Work with a partner.

a. Tell whether the new designs are proportional to the original design. Explain your reasoning.

Original 

Design 1 

Design 2 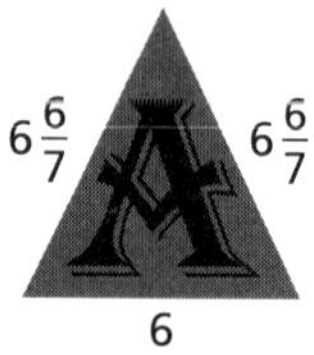

b. Draw two designs that are proportional to the given design. Make one bigger and one smaller. Label the sides of the designs with their lengths.

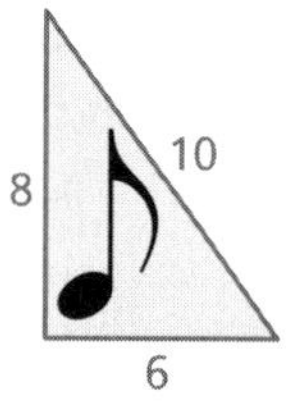

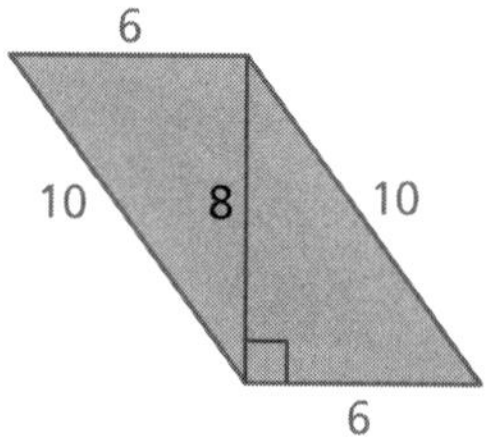

What Is Your Answer?

3. **IN YOUR OWN WORDS** How can you use proportions to help make decisions in art, design, and magazine layouts? Give two examples.

4. **a.** Use a computer art program to draw two rectangles that are proportional to each other.

   b. Print the two rectangles on the same piece of paper.

   c. Use a centimeter ruler to measure the length and width of each rectangle. Record your measurements here.

"I love this statue. It seems similar to a big statue I saw in New York."

   d. Find the following ratios. What can you conclude?

$$\frac{\text{Length of Larger}}{\text{Length of Smaller}} \qquad \frac{\text{Width of Larger}}{\text{Width of Smaller}}$$

Name ______________________________ Date __________

6.2 Practice
For use after Lesson 6.2

Name the corresponding angles and the corresponding sides of the similar figures.

1. 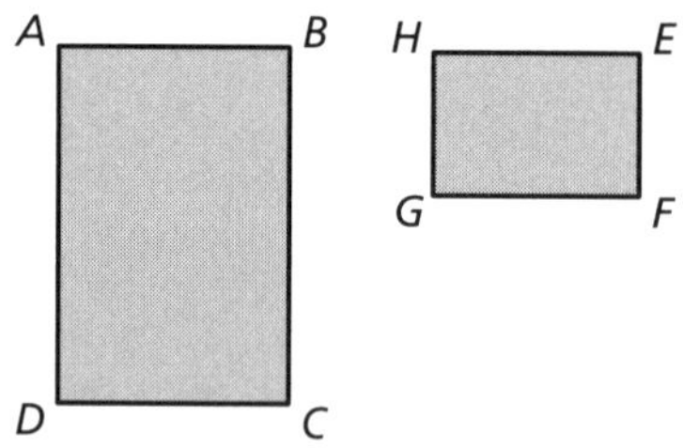

2. 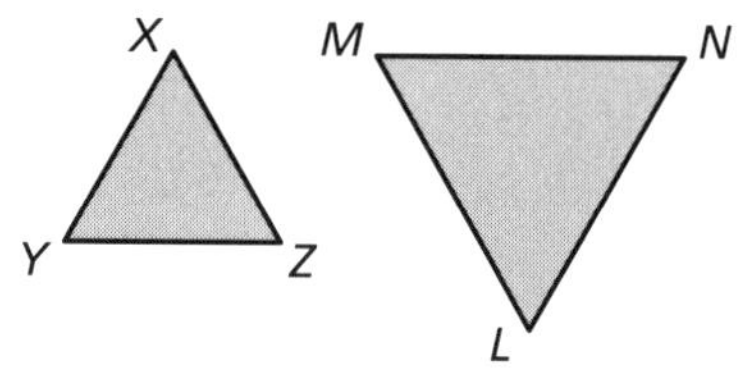

Tell whether the two figures are similar. Explain your reasoning.

3. 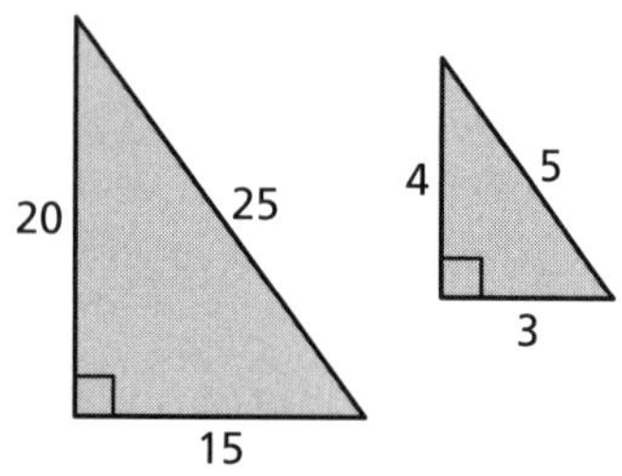

4. 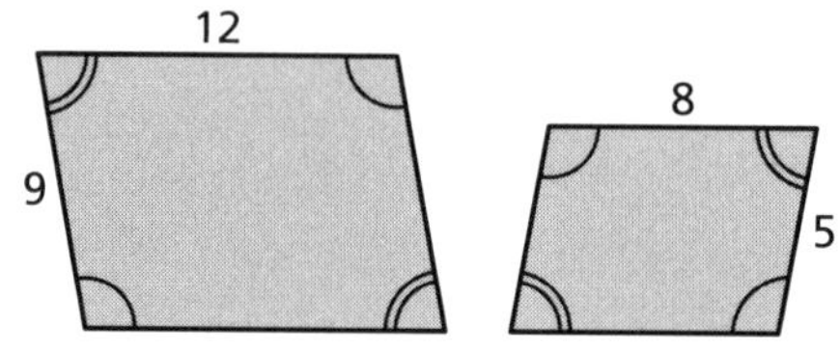

5. In your classroom, a dry erase board is 8 feet long and 4 feet wide. Your teacher makes individual dry erase boards for you to use at your desk that are 11.5 inches long and 9.5 inches wide. Are the boards similar?

6. You have a 4 x 6 photo of you and your friend.

a. You order a 5 x 7 print of the photo. Is the new photo similar to the original?

b. You enlarge the original photo to three times its size on your computer. Is the new photo similar to the original?

Name______________________________ Date__________

6.3 Perimeters and Areas of Similar Figures

For use with Activity 6.3

Essential Question How do changes in dimensions of similar geometric figures affect the perimeters and areas of the figures?

1 ACTIVITY: Comparing Perimeters and Areas

Work with a partner. Use pattern blocks to make a figure whose dimensions are 2, 3, and 4 times greater than those of the original figure. Find the perimeter *P* and area *A* of each larger figure.

a. Square

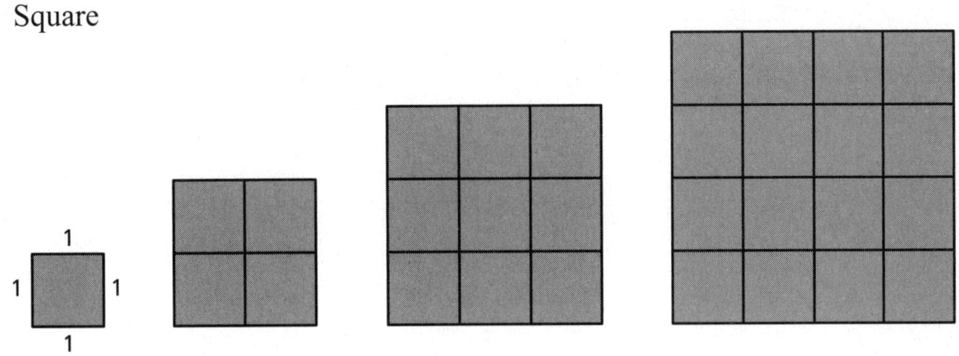

$P = 4$ $P =$ ______ $P =$ ______ $P =$ ______

$A = 1$ $A =$ ______ $A =$ ______ $A =$ ______

b. Triangle

$P = 3$ $P =$ ______ $P =$ ______ $P =$ ______

$A = B$ $A =$ ______ $A =$ ______ $A =$ ______

Name ______________________________ Date __________

c. Rectangle

$P = 6$

$A = 2$

d. Parallelogram

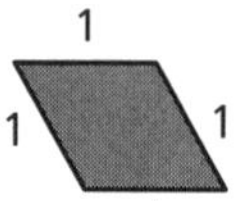

$P = 4$

$A = C$

2 ACTIVITY: Finding Patterns for Perimeters

Work with a partner. Complete the table for the perimeters of the figures in Activity 1. Describe the pattern.

Figure	Original Side Lengths	Double Side Lengths	Triple Side Lengths	Quadruple Side Lengths
	$P = 4$			
	$P = 3$			
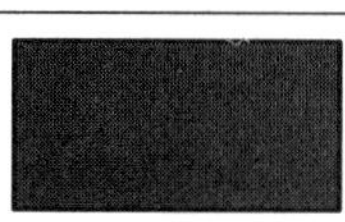	$P = 6$			
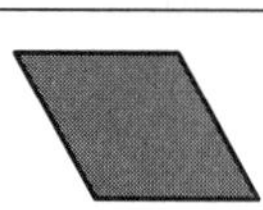	$P = 4$			

Name____ Date__________

6.3 Perimeters and Areas of Similar Figures (continued)

3 ACTIVITY: Finding Patterns for Areas

Work with a partner. Complete the table for the areas of the figures in Activity 1. Describe the pattern.

Figure	Original Side Lengths	Double Side Lengths	Triple Side Lengths	Quadruple Side Lengths
	$A = 1$			
	$A = B$			
	$A = 2$			
	$A = C$			

What Is Your Answer?

4. **IN YOUR OWN WORDS** How do changes in dimensions of similar geometric figures affect the perimeters and areas of the figures?

Name ________________________________ Date __________

6.3 Practice

For use after Lesson 6.3

The two figures are similar. Find the ratios (shaded to nonshaded) of the perimeters and of the areas.

1.

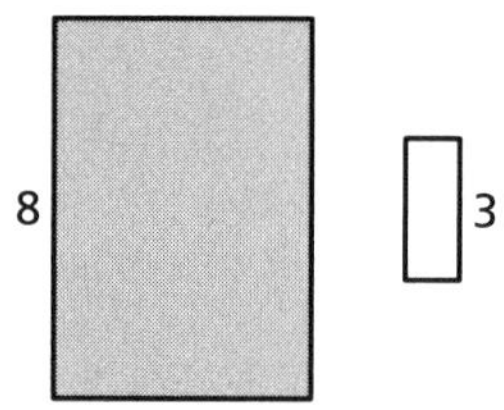

2.

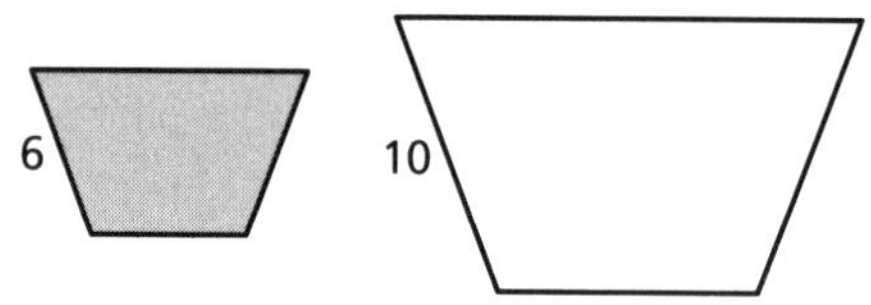

3.

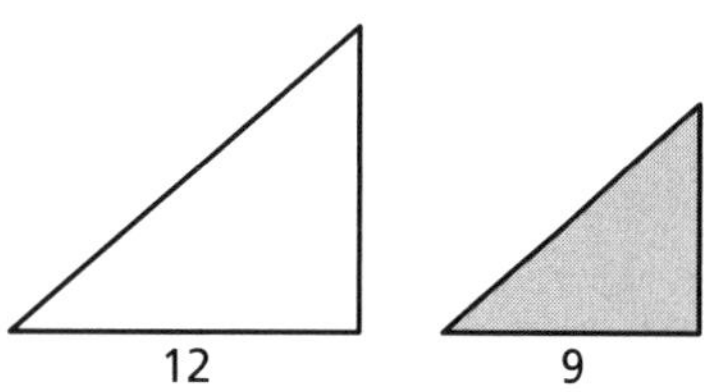

4.

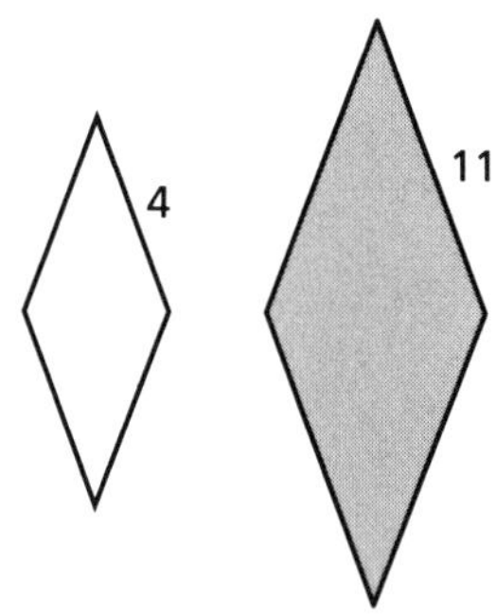

5. You buy two picture frames that are similar. The ratio of the corresponding side lengths is 4 : 5. What is the ratio of the areas?

6. Rectangle A is similar to Rectangle B. What is the ratio of the perimeter of Rectangle A to the perimeter of Rectangle B?

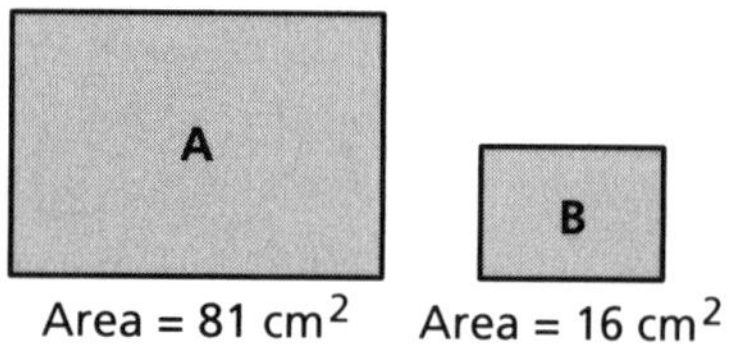

Name____________________ Date__________

6.4 Finding Unknown Measures in Similar Figures

For use with Activity 6.4

Essential Question What information do you need to know to find the dimensions of a figure that is similar to another figure?

1 ACTIVITY: Drawing and Labeling Similar Figures

Work with a partner. You are given the rectangle. Find another rectangle that is similar and has one side from $(-1, -6)$ to $(5, -6)$. Label the vertices.

a.

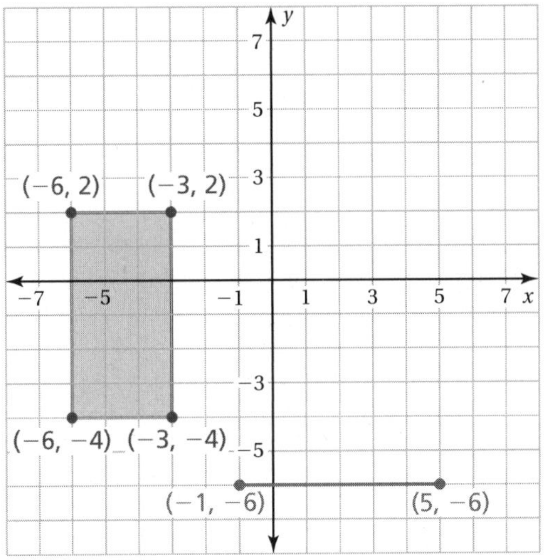

You can see that the two rectangles are similar by showing that ratios of corresponding sides are equal. Complete the steps below. Are the two rectangles above similar?

$$\frac{\text{Shaded Length}}{\text{Unshaded Length}} \stackrel{?}{=} \frac{\text{Shaded Width}}{\text{Unshaded Length}}$$

$$\frac{\text{change in } y}{\text{change in } y} \stackrel{?}{=} \frac{\text{change in } x}{\text{change in } x}$$

$$\frac{\square}{\square} = \frac{\square}{\square}$$

6.4 Finding Unknown Measures in Similar Figures (continued)

b. There are three other rectangles that are similar to the shaded rectangle and have the given side.

- Draw each one. Label the vertices of each.

- Show that each is similar to the original shaded rectangle.

2 ACTIVITY: Reading a Map

Work with a partner.

a. The rectangles are similar. Find the length of the larger rectangle. Explain your reasoning.

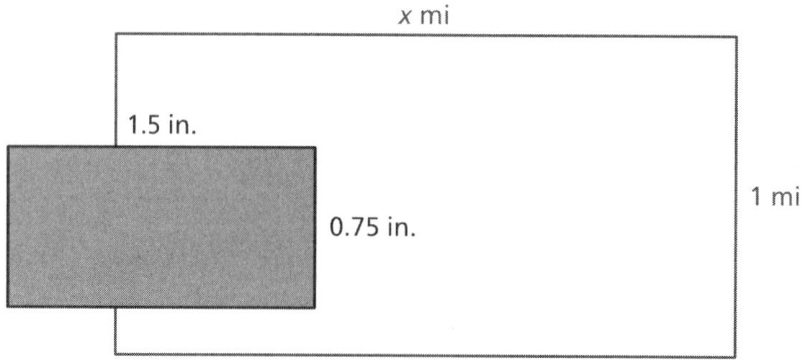

b. The distance marked by the vertical line on the map is 1 mile. Find the distance marked by the horizontal line. Explain your reasoning.

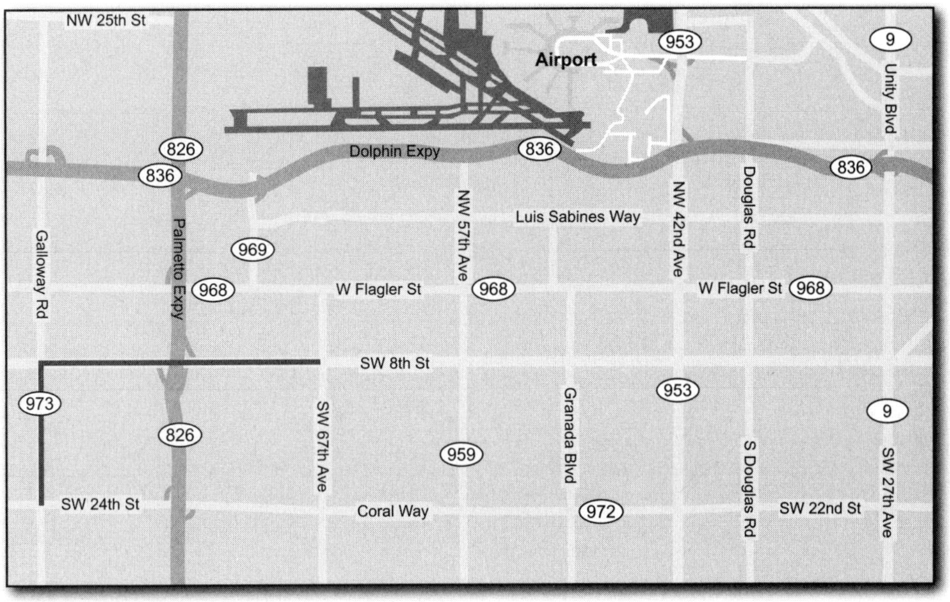

What Is Your Answer?

3. IN YOUR OWN WORDS What information do you need to know to find the dimensions of a figure that is similar to another figure? Give some examples using two rectangles.

4. When you know the length and width of one rectangle and the length of a similar rectangle, can you always find the missing width? Why or why not?

Name ______________________ Date __________

6.4 Practice
For use after Lesson 6.4

The polygons are similar. Find the value of *x*.

1.

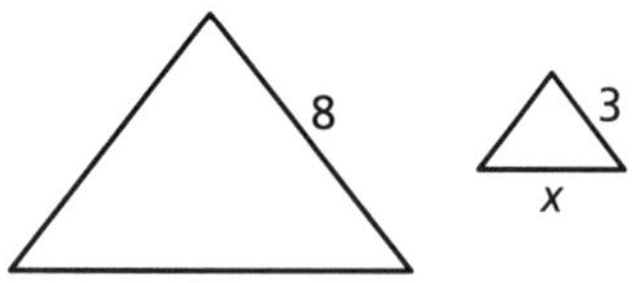

2.

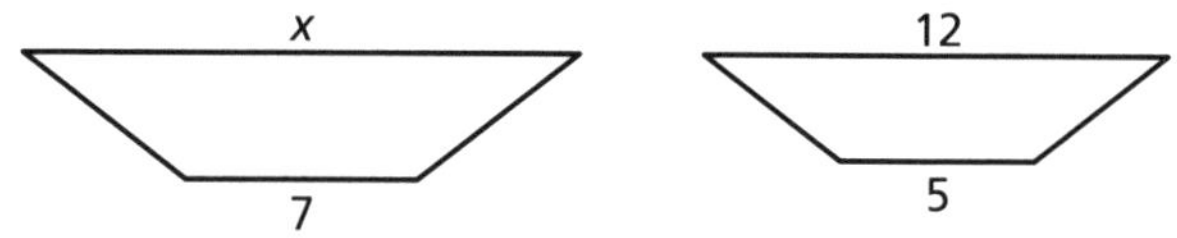

3.

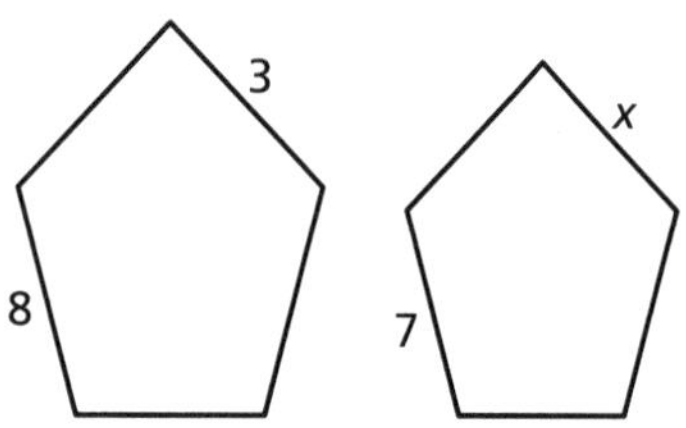

4. The ratio of the perimeters is 2 : 5.

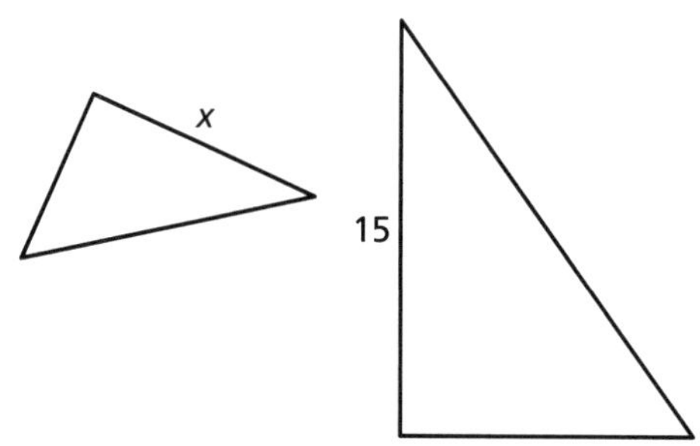

5. A tree casts a shadow that is 50 feet long. A 4-foot person casts a shadow that is 10 feet long. How tall is the tree?

6. A cookie sheet is 12 inches wide and has a perimeter of 52 inches. You buy a similar cookie sheet that is 15 inches wide. What is its perimeter?

Name____ Date__________

6.5 Scale Drawings

For use with Activity 6.5

Essential Question How can you use a scale drawing to estimate the cost of painting a room?

1 ACTIVITY: Making Scale Drawings

Work with a partner. You have decided that your classroom needs to be painted. Start by making a scale drawing of each of the four walls.

- **Measure each of the walls.**
- **Measure the locations and dimensions of parts that will *not* be painted.**
- **Decide on a scale for your drawings.**
- **Make a scale drawing of each of the walls.**

Sample: Wall #1

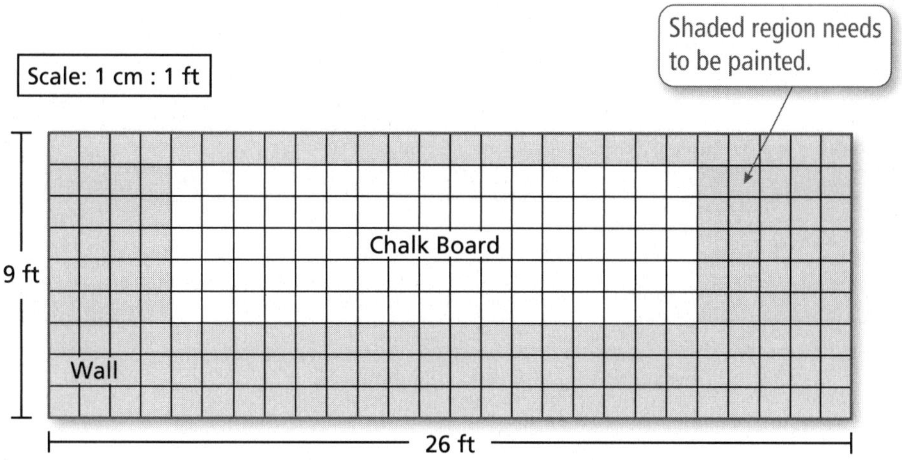

- **For each wall, find the area of the part that needs to be painted.**

	Dimensions	*Area*
Dimensions of the wall	9 ft by 26 ft	$9 \times 26 = 234$ sq ft
Dimensions of the part that will *not* be painted	5 ft by 17 ft	$5 \times 17 = 85$ sq ft
Area of painted part		$234 - 85 = 149$ sq ft

Name ______________________________ Date __________

6.5 Scale Drawings (continued)

2 ACTIVITY: Using Scale Drawings

Work with a partner.

You are using a paint that covers 200 square feet per gallon. Each wall will need two coats of paint.

Interior latex paint	$40 per gallon
Roller, pan, and brush set	$12

a. Find the total area of the walls in your classroom that need to be painted.

b. Find the amount of paint you need to buy.

c. Estimate the total cost of painting your classroom.

Name____ Date__________

What Is Your Answer?

3. **IN YOUR OWN WORDS** How can you use a scale drawing to estimate the cost of painting a room?

4. Use a scale drawing to estimate the cost of painting another room, such as your bedroom or another room in your house.

5. Look at some maps in your school library or on the Internet. Make a list of the different scales used on the maps.

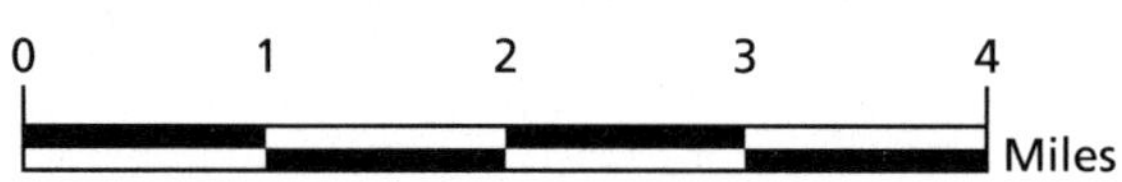

"I don't get it. According to this map, we only have to drive $8\frac{1}{2}$ inches."

6. When you view a map on the Internet, how does the scale change when you zoom out? How does the scale change when you zoom in?

Name ____ Date __________

6.5 Practice

For use after Lesson 6.5

Find the missing dimension. Use the scale factor 1 : 8.

Item	Model	Actual
1. Statue	Height: 168 in.	Height: ________ ft
2. Painting	Width: ________ cm	Width: 200 m
3. Alligator	Height: ________ in.	Height: 6.4 ft
4. Train	Length: 36.5 in.	Length: ________ ft

5. The diameter of the moon is 2160 miles. A model has a scale of 1 in. : 150 mi. What is the diameter of the model?

6. A map has a scale of 1 in. : 4 mi.

a. You measure 3 inches between your house and the movie theater. How many miles is it from your house to the movie theater?

b. It is 17 miles to the mall. How many inches is that on the map?

Name____ Date__________

Chapter 7 Fair Game Review

How many lines of symmetry does the figure have?

1.

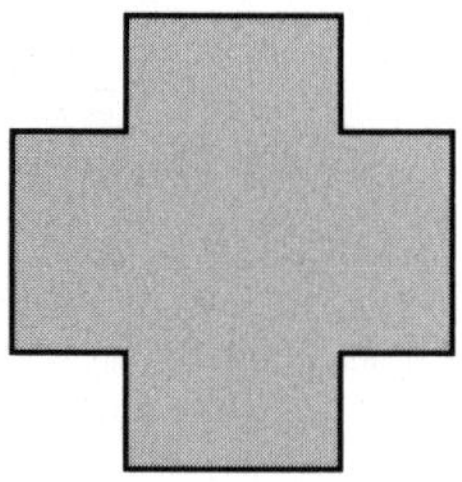

2.

3.

4.

5.

6.

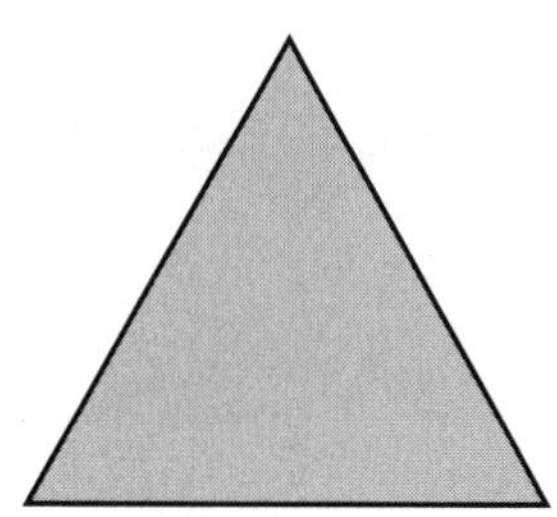

7. Draw the lines of symmetry on the square. How many lines of symmetry does a square have?

Name ______________________ Date __________

Chapter 7 Fair Game Review (continued)

Tell whether the figure has rotational symmetry.

8. 

9. 

10. 

11. 

12. 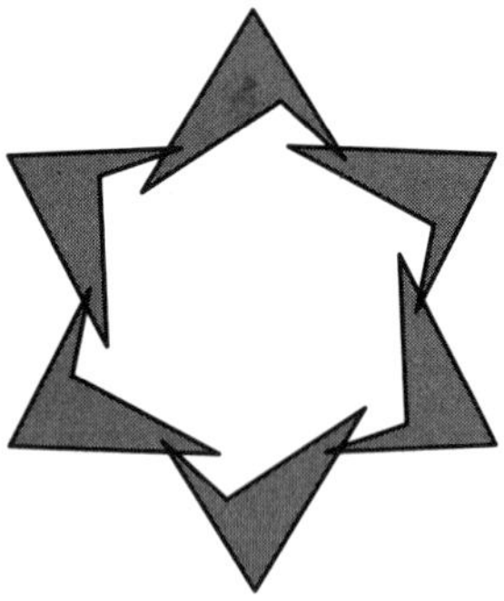

13. 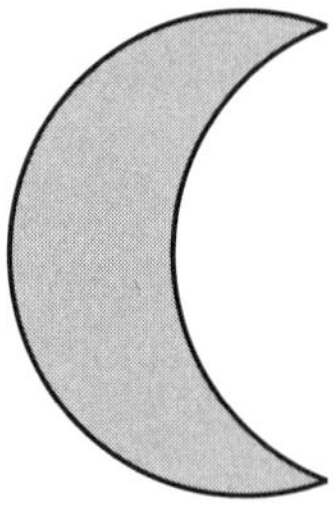

14. You make a trapezoidal photo collage. Does the collage have rotational symmetry?

Name______________________________ Date__________

7.1 Translations

For use with Activity 7.1

Essential Question How can you use translations to make a tessellation?

When you slide a tile it is called a **translation**. When tiles can be used to cover a floor with no empty spaces, the collection of tiles is called a *tessellation*.

1 ACTIVITY: Describing Tessellations

Work with a partner. Can you make the pattern by using a translation of single tiles that are all of the same shape and design? If so, outline the single tile.

a. Sample:

Tile Pattern

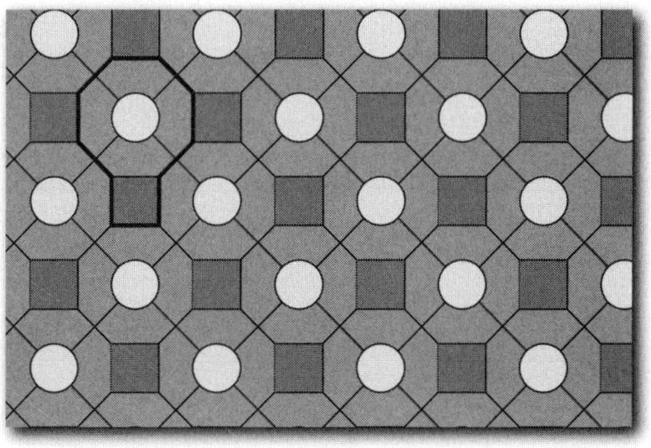

Single Tiles

b.

c.

7.1 Translations (continued)

d. 

e. 

2 ACTIVITY: Tessellations and Basic Shapes

Work with a partner.

a. Which pattern blocks can you use to make a tessellation?

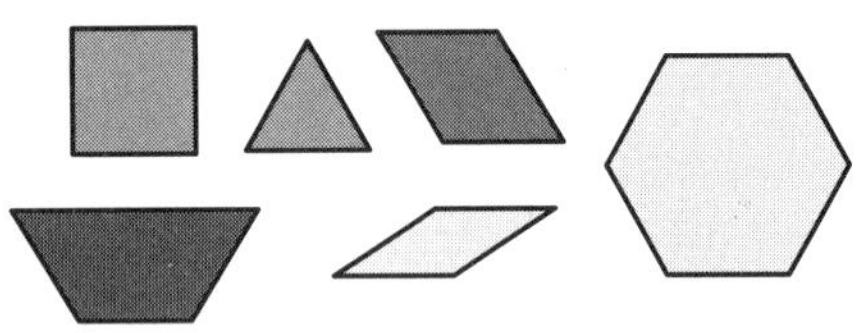

b. For each pattern block you listed in part (a), draw the tessellation.

c. Can you make the tessellation using only translations, or do you have to rotate or flip the pattern blocks?

7.1 Translations (continued)

3 ACTIVITY: Designing Tessellations

Work with a partner. Design your own tessellation. Use one of the basic shapes from Activity 2.

Sample:

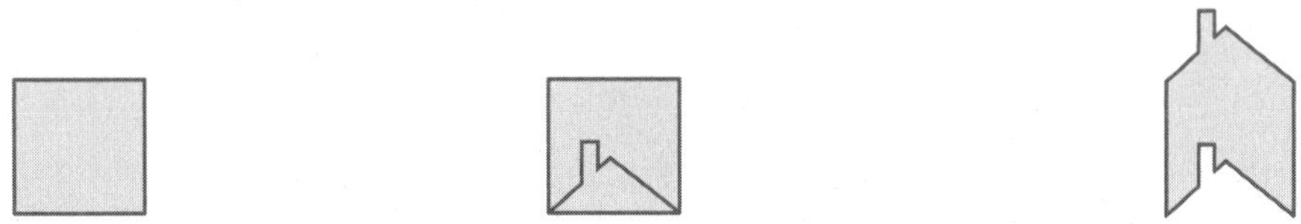

Start with a square. | Cut a design out of one side. | Tape it to the other side to make your pattern.

- **Use your pattern and translations to make your tessellation.**
- **Color the tessellation.**

What Is Your Answer?

4. **IN YOUR OWN WORDS** How can you use translations to make a tessellation? Give an example.

5. Draw any parallelogram. Does it tessellate? Is it true that any parallelogram can be translated to make a tessellation? Explain why.

Name ______________________________ Date __________

7.1 Practice

For use after Lesson 7.1

Tell whether the shaded figure is a translation of the nonshaded figure.

1. 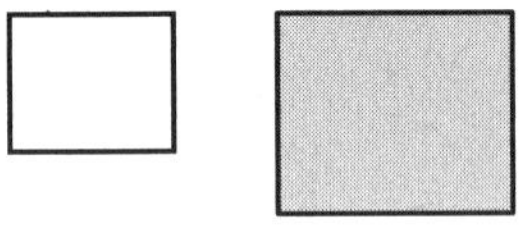

2. 

3. 

4. Translate the square 4 units left and 1 unit down. What are the coordinates of the image?

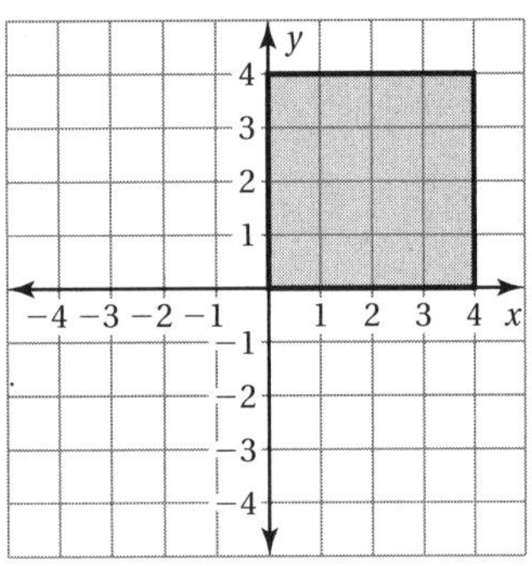

5. Translate the triangle 5 units right and 4 units up. What are the coordinates of the image?

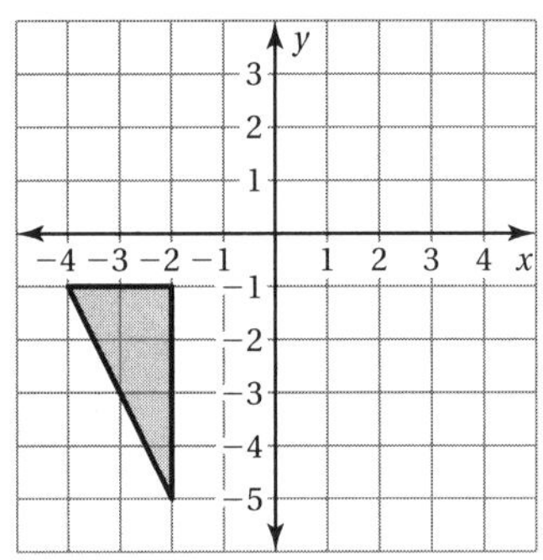

6. A rectangle is translated 3 units left and 4 units down. Then the image is translated 6 units right and 1 unit up. Write a translation of the original rectangle to the ending position.

7. Describe the translation from the shaded figure to the nonshaded figure.

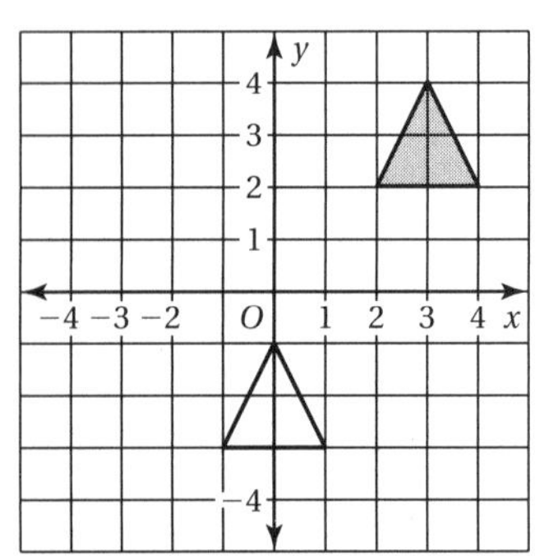

Name____ Date__________

7.2 Reflections

For use with Activity 7.2

Essential Question How can you use reflections to classify a frieze pattern?

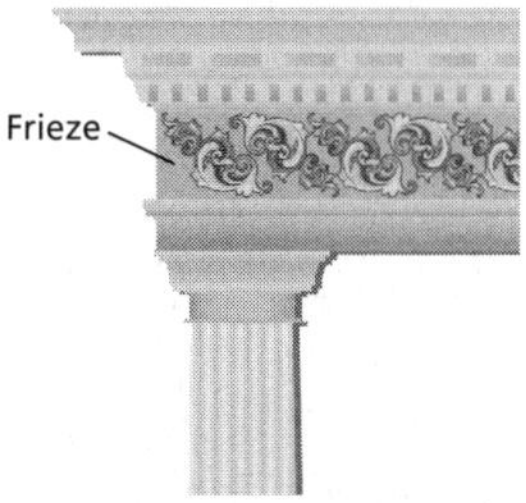

A *frieze* is a horizontal band that runs at the top of a building. A frieze is often decorated with a design that repeats.

- All frieze patterns are translations of themselves.
- Some frieze patterns are reflections of themselves.

1 EXAMPLE: Frieze Patterns

Is the frieze pattern a reflection of itself when folded horizontally? vertically?*

- Fold (reflect) on the horizontal axis. The pattern coincides.

- Fold (reflect) on the vertical axis. The pattern coincides.

The frieze pattern is a reflection of itself when it is folded horizontally *and* vertically.

2 ACTIVITY: Frieze Patterns and Reflections

Work with a partner. Is the frieze pattern a reflection of itself when folded *horizontally*, *vertically*, or *neither*?

a.

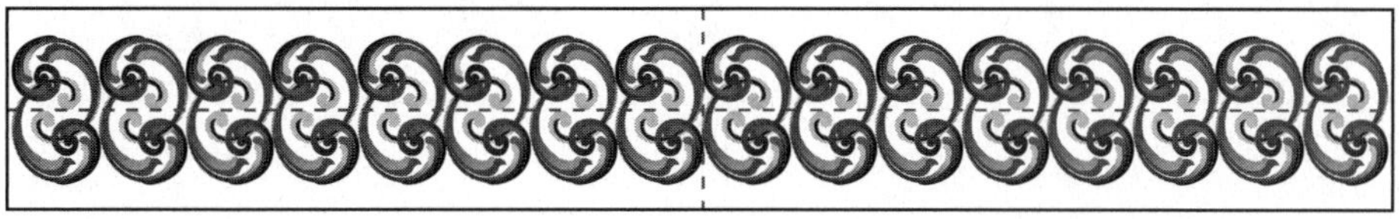

*Cut-outs are available in the back of the Record and Practice Journal.

Name ____ Date __________

b.

c.

d.

e.

f.

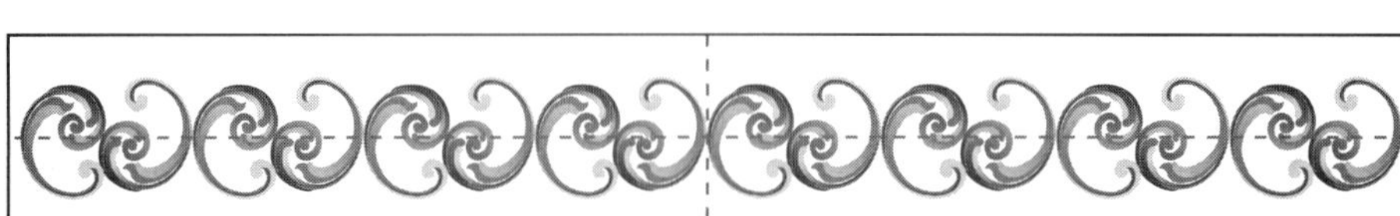

Name____ Date__________

7.2 **Reflections (continued)**

What Is Your Answer?

3. Draw a frieze pattern that is a reflection of itself when folded horizontally.

4. Draw a frieze pattern that is a reflection of itself when folded vertically.

5. Draw a frieze pattern that is not a reflection of itself when folded horizontally or vertically.

6. **IN YOUR OWN WORDS** How can you use reflections to classify a frieze pattern?

Name ______________________________ Date __________

7.2 Practice

For use after Lesson 7.2

Tell whether the shaded figure is a reflection of the nonshaded figure.

1. 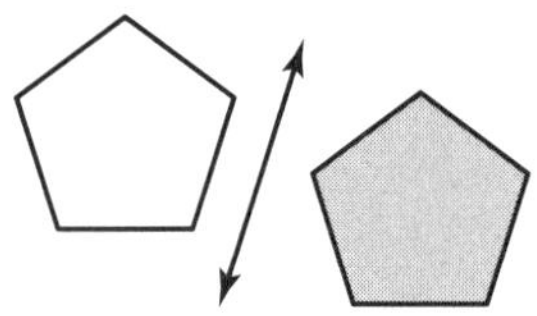

2. 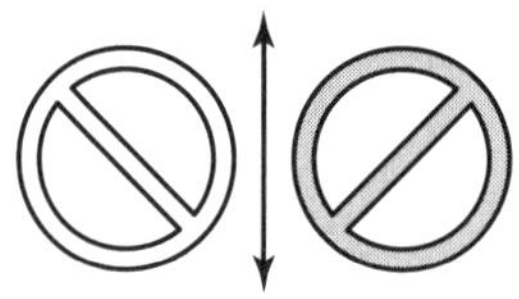

3. 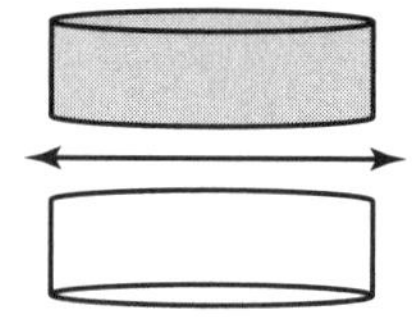

Find the coordinates of the figure after reflecting in the *x*-axis.

4. $A(1, 2)$, $B(3, 2)$, $C(1, 4)$

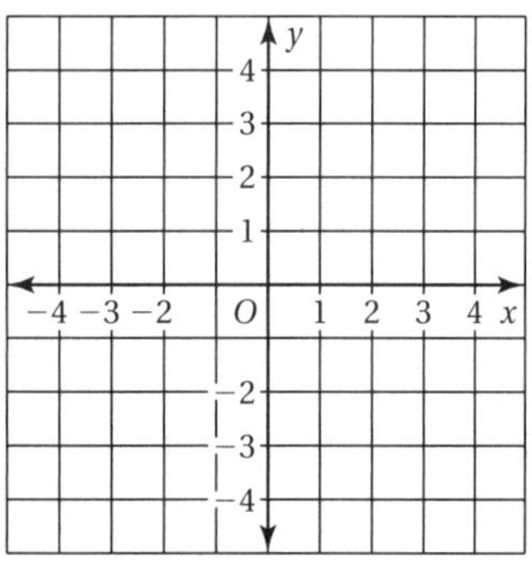

5. $W(3, 1)$, $X(3, 4)$, $Y(1, 4)$, $Z(1, 1)$

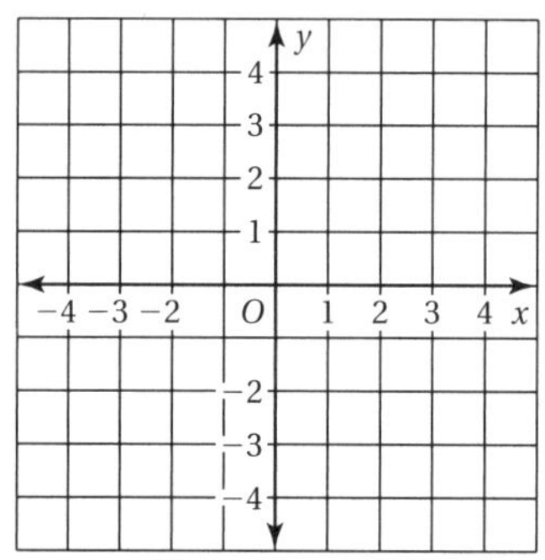

Find the coordinates of the figure after reflecting in the *y*-axis.

6. $J(3, 4)$, $K(3, 0)$, $L(2, 4)$

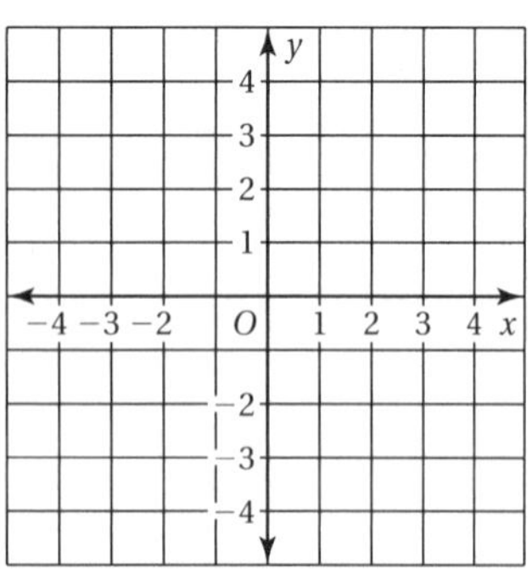

7. $M(2, 2)$, $N(2, 3)$, $P(3, 3)$, $Q(4, 1)$

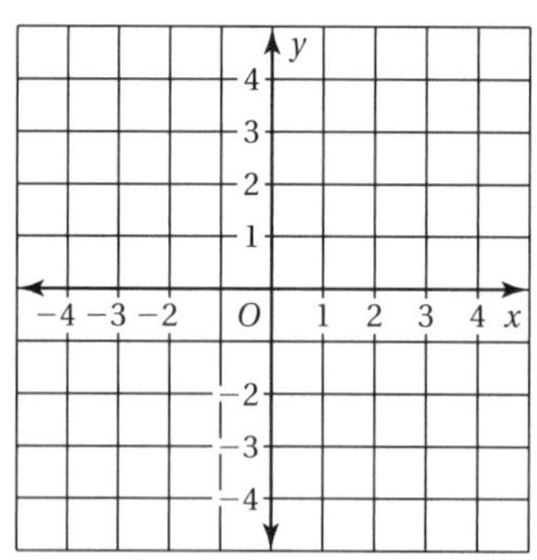

8. In a pinball game, if you perfectly reflect the ball off of the wall, will the ball hit the bonus target?

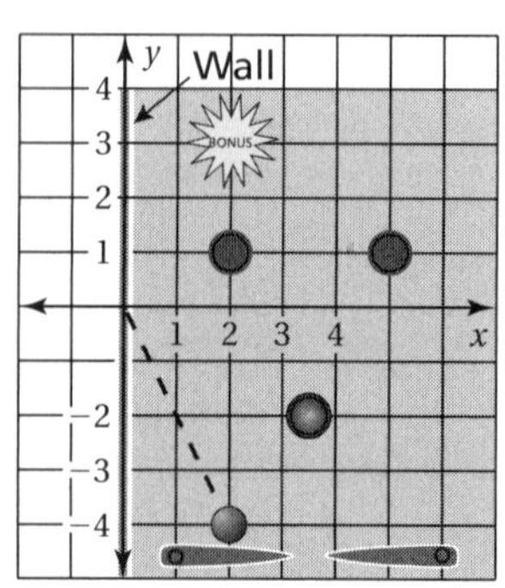

Name____ Date__________

7.3 Rotations

For use with Activity 7.3

Essential Question What are the three basic ways to move an object in a plane?

1 ACTIVITY: Three Basic Ways to Move Things

There are three basic ways to move objects on a flat surface.

1. Translate the object. **2.** Reflect the object. **3.** Rotate the object.

Work with a partner.

- Cut out a paper triangle that is the same size as the shaded triangle shown.*
- Decide how you can move the shaded triangle to make each nonshaded triangle.
- Is each move a *translation*, a *reflection*, or a *rotation*?

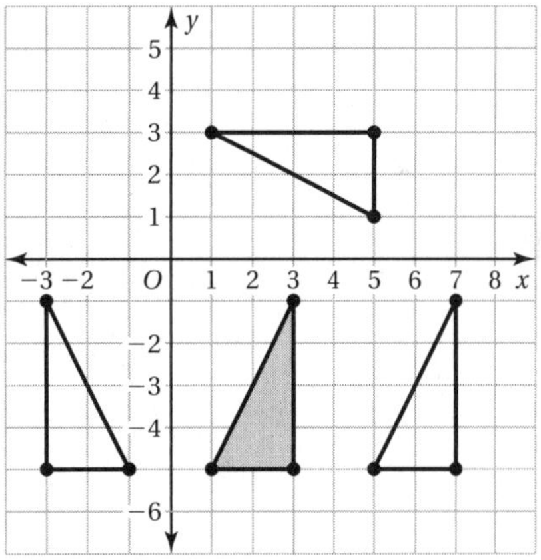

- Draw four other nonshaded triangles in a coordinate plane. Describe how you can move the shaded triangle to make each nonshaded triangle.

*Cut-outs are available in the back of the Record and Practice Journal.

Name ______________________________ Date ________

2 ACTIVITY: Tessellating a Plane

Work with a partner.

a. Describe how the figure labeled 1 in each diagram can be moved to make the other figures.

Triangles

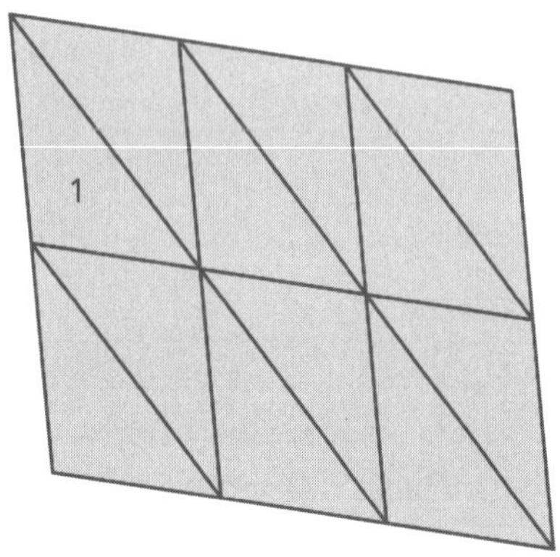

Quadrilaterals

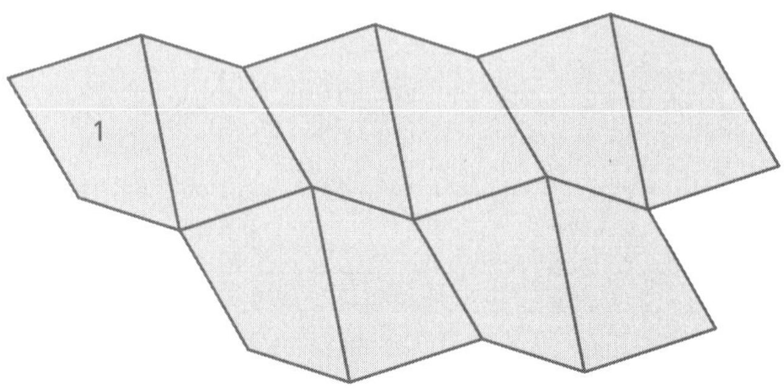

b. EXPERIMENT Will *any* triangle tessellate? Conduct an experiment to gather information to help form your conclusion. Draw a triangle. Cut it out. Then use it to trace other triangles so that you cover the plane with triangles that are all the same shape.

c. EXPERIMENT Will *any* quadrilateral tessellate? Conduct an experiment to gather information to help form your conclusion. Draw a quadrilateral. Cut it out. Then use it to trace other quadrilaterals so that you cover the plane with quadrilaterals that are all the same shape.

What Is Your Answer?

3. IN YOUR OWN WORDS What are the three basic ways to move an object in a plane? Draw an example of each.

"Dear Sub Shop: Why do you put the cheese on the subs so some parts have double coverage and some have none?"

"My suggestion is that you use the tessellation property of triangles for even cheese coverage."

Name ______________________________ Date __________

7.3 Practice

For use after Lesson 7.3

Tell whether the shaded figure is a rotation of the nonshaded figure about the origin. If so, give the angle and the direction of rotation.

1.

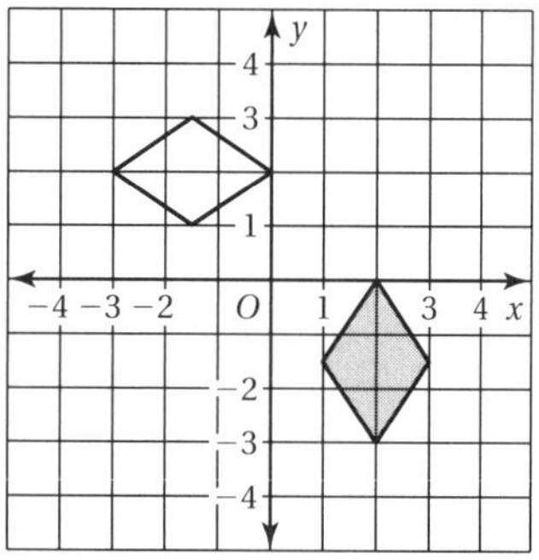

2.

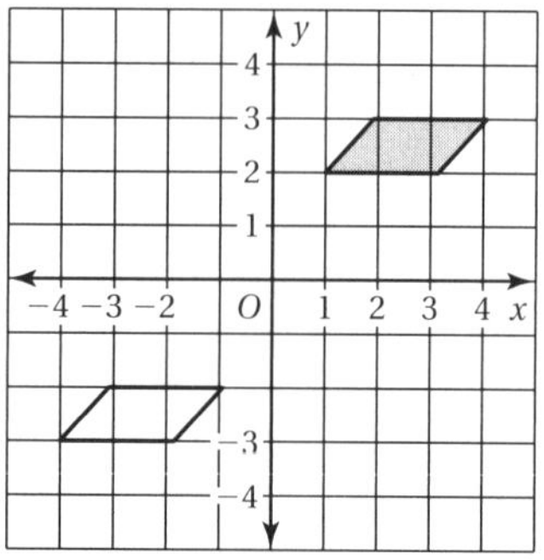

The vertices of a triangle are $A(1, 1)$, $B(3, 1)$, and $C(3, 4)$. Rotate the triangle about the origin as described. Find the coordinates of the image.

3. 90° clockwise

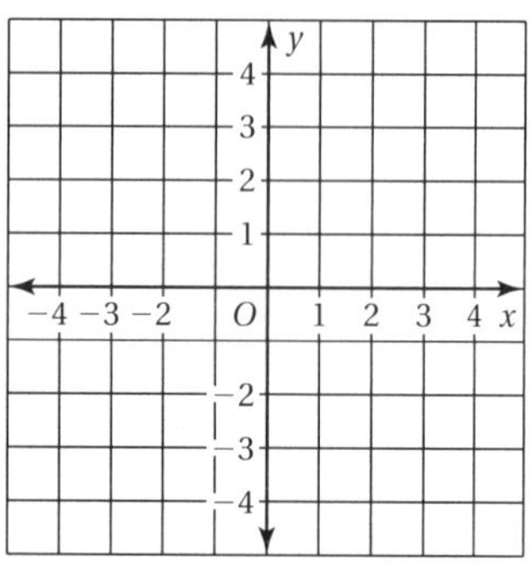

4. 180°

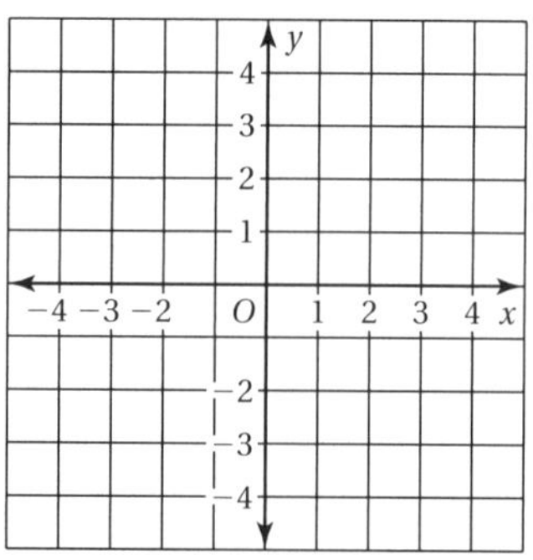

5. A triangle is rotated 180° about the origin. Its image is reflected in the x-axis. The vertices of the final triangle are $(-4, -4)$, $(-2, -4)$, and $(-3, -1)$. What are the vertices of the original triangle?

Name____ Date__________

7.4 Dilations

For use with Activity 7.4

Essential Question How can you enlarge or reduce a polygon in the coordinate plane?

1 ACTIVITY: Comparing Triangles in a Coordinate Plane

Work with a partner. Write the coordinates of the vertices of the shaded triangle. Then write the coordinates of the vertices of the nonshaded triangle.

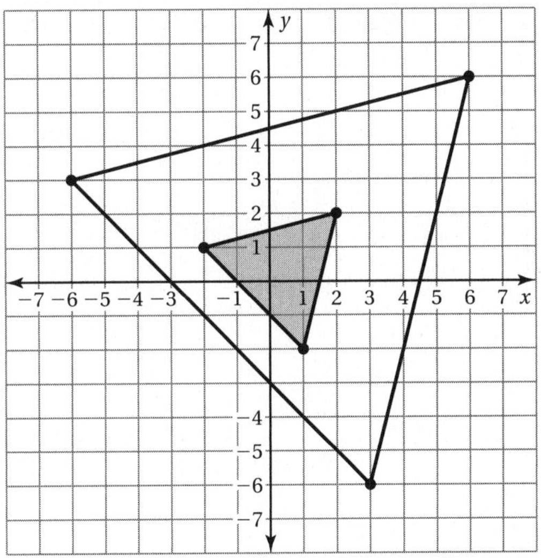

a. How are the two sets of coordinates related?

b. How are the two triangles related? Explain your reasoning.

c. Draw a dashed triangle whose coordinates are twice the values of the corresponding coordinates of the shaded triangle. How are the dashed and shaded triangles related? Explain your reasoning.

7.4 Dilations (continued)

d. How are the coordinates of the nonshaded and dashed triangles related? How are the two triangles related? Explain your reasoning.

2 ACTIVITY: Drawing Triangles in a Coordinate Plane

Work with a partner.

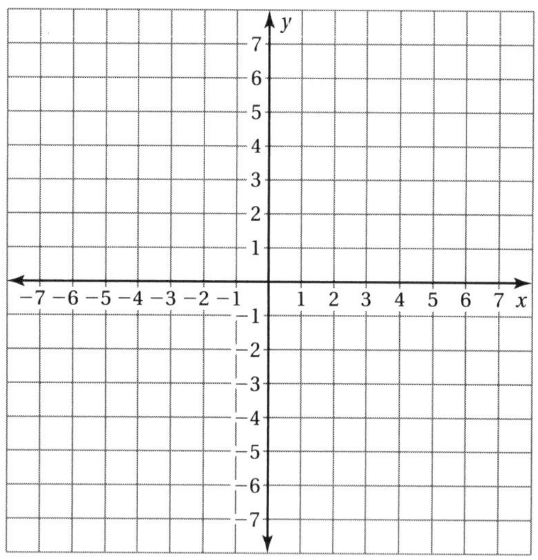

a. Draw the triangle whose vertices are $(0, 2)$, $(-2, 2)$, and $(1, -2)$.

b. Multiply each coordinate of the vertices by 2 to obtain three new vertices. Draw the triangle given by the three new vertices. How are the two triangles related?

c. Repeat part (b) by multiplying by 3 instead of 2.

7.4 Dilations (continued)

3 ACTIVITY: Summarizing Transformations

Work with a partner. Make a table that summarizes the relationships between the original figure and its image for the 4 types of transformations you studied in this chapter.

What Is Your Answer?

4. **IN YOUR OWN WORDS** How can you enlarge or reduce a polygon in the coordinate plane?

5. Describe how knowing how to enlarge or reduce figures in a technical drawing is important in a career such as drafting.

Name ____________________ Date __________

7.4 Practice

For use after Lesson 7.4

Tell whether the shaded figure is a dilation of the nonshaded figure.

1.

2.

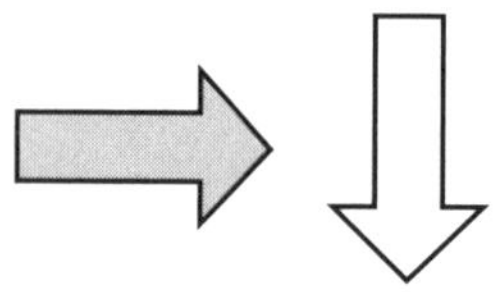

3.

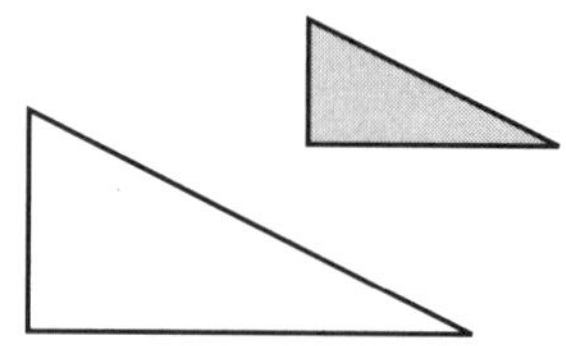

The vertices of a polygon are given. Draw the polygon and its image after a dilation with the given scale factor. Identify the type of dilation.

4. $A(-2, 2), B(1, 2), C(1, -1); k = 3$

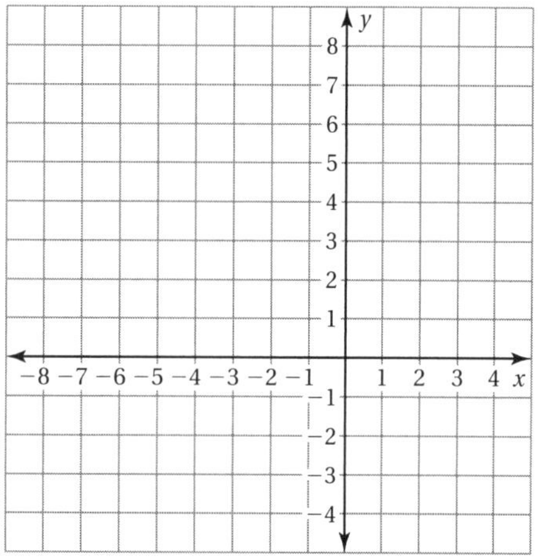

5. $D(4, 2), E(4, 8), F(8, 8), G(8, 2); k = \frac{1}{2}$

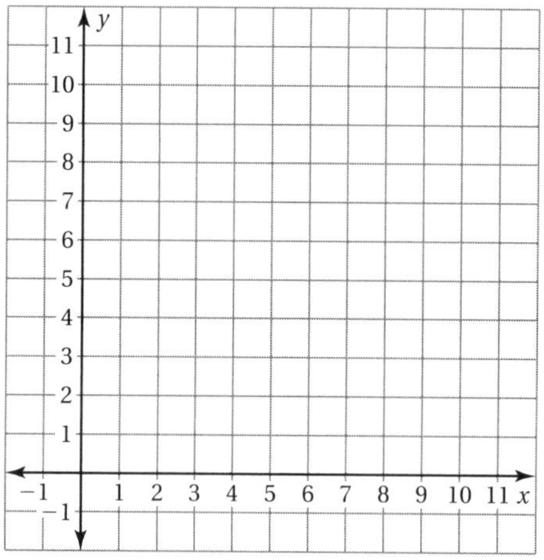

6. A rectangle is dilated using a scale factor of 6. The image is then dilated using a scale factor of $\frac{1}{3}$. What scale factor could you use to dilate the original rectangle to get the final rectangle? Explain.

Name____________________________________ Date__________

Chapter 8 Fair Game Review

Find the area of the square or rectangle.

1.

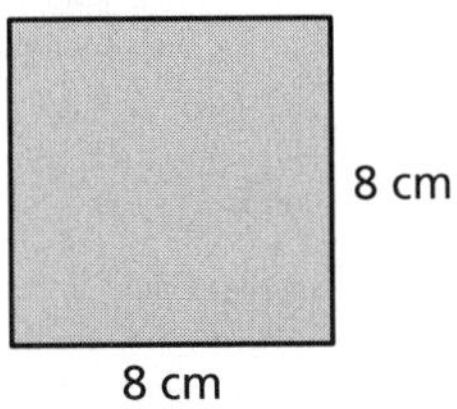

2.

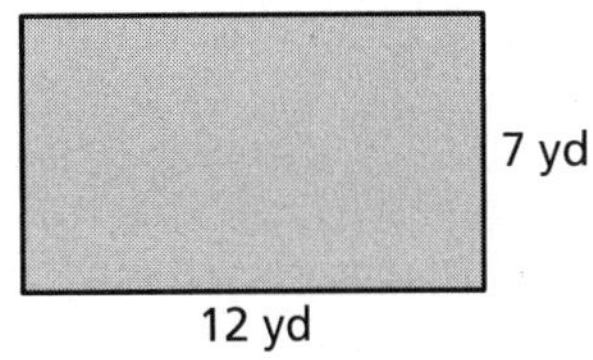

3.

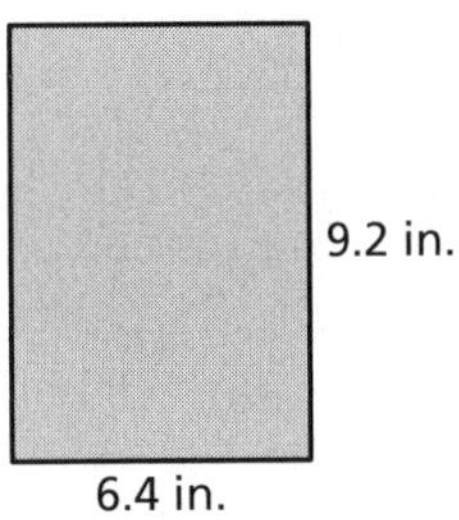

4.

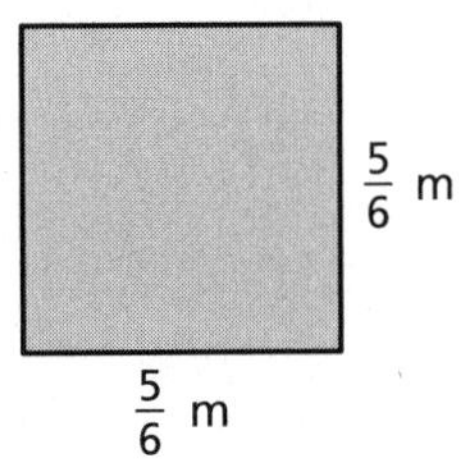

5.

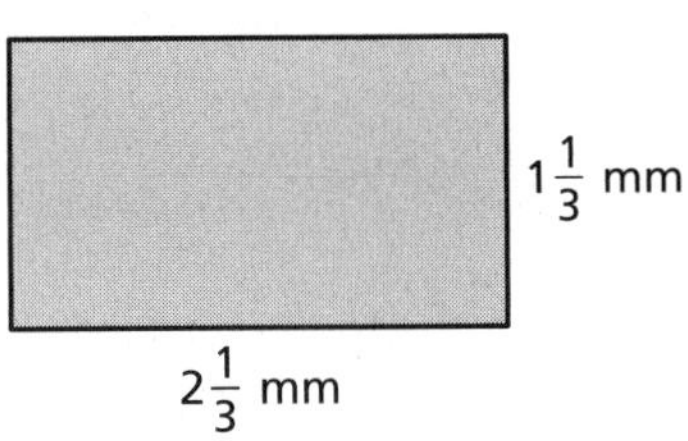

6.

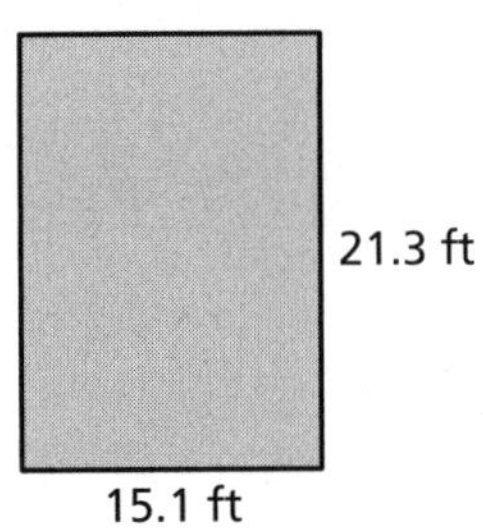

7. An artist buys a square canvas with a side length of 2.5 feet. What is the area of the canvas?

Name ______________________ Date __________

Chapter 8 Fair Game Review (continued)

Find the area of the circle. Use 3.14 or $\frac{22}{7}$ for π.

8.

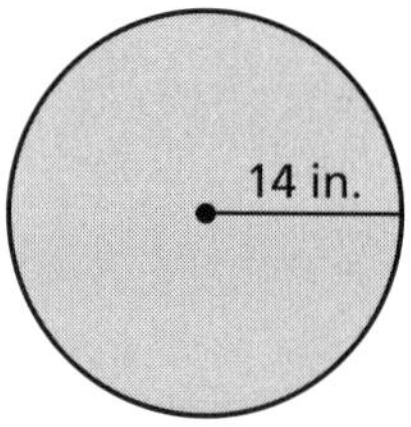

9.

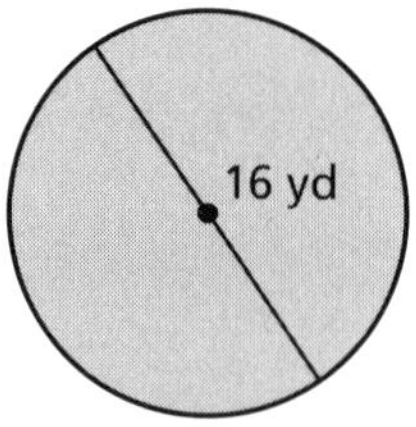

10.

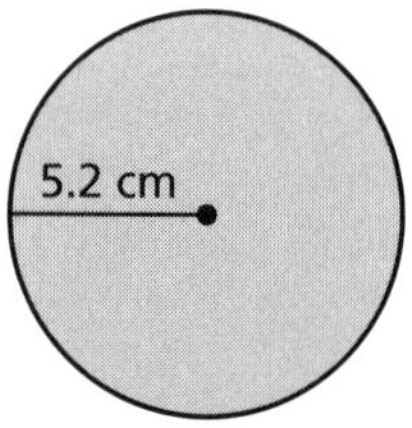

11.

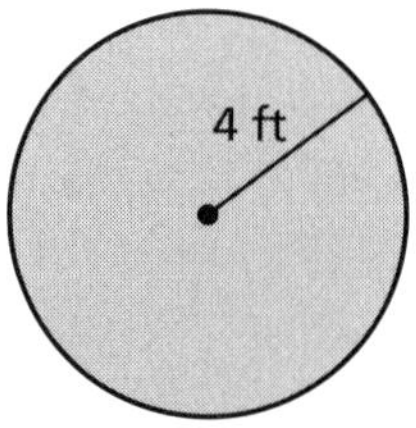

12.

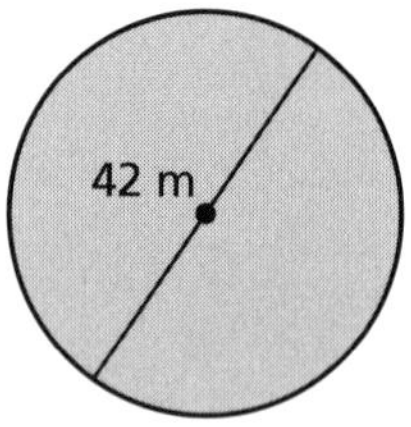

13.

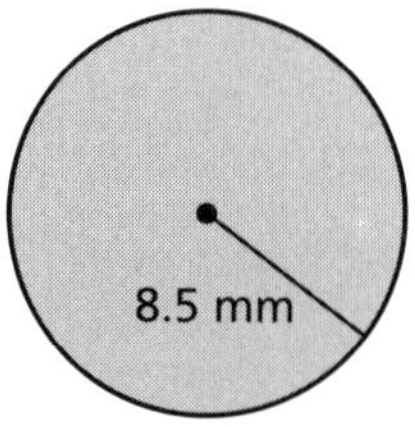

14. The face of a grandfather clock has a diameter of 3 feet. What is the area of the face of the clock? Use 3.14 for π.

Name____________________ Date__________

8.1 Surface Areas of Rectangular Prisms

For use with Activity 8.1

Essential Question How does changing a dimension of a rectangular prism affect its surface area?

1 ACTIVITY: Comparing Surface Areas

Work with a partner.

a. Copy the net shown below on grid paper.

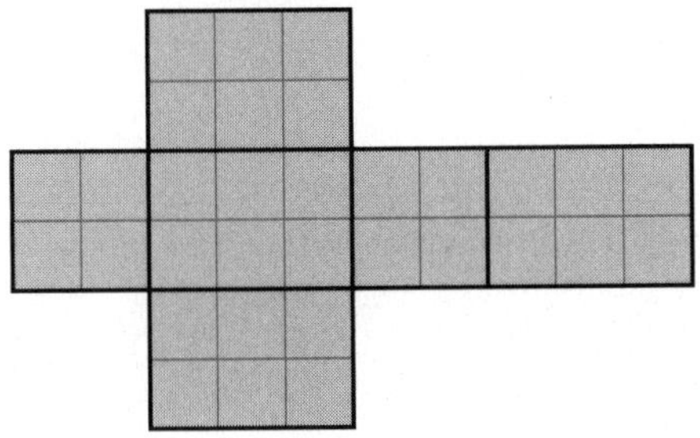

b. Double the width of the rectangular prism whose net is shown. Draw a net for the new prism.

c. Cut out both nets and fold them to form two prisms. Compare the two prisms visually. Estimate how much more surface area the larger prism has than the smaller prism.

d. Find the actual surface area of each prism. How close was your estimate?

Note: Save your nets from Activities 1 and 2. You will use them again in Activity 8.2.

8.1 Surface Areas of Rectangular Prisms (continued)

2 ACTIVITY: Comparing Surface Areas

Work with a partner.

a. Double all three dimensions of the original prism in Activity 1: the width, length, and height.

New width: ________________

New length: ________________

New height: ________________

b. Calculate the surface area of the new prism. How many times greater is the surface area of the new prism?

c. Triple all three dimensions of the original prism in Activity 1.

New width: ________________

New length: ________________

New height: ________________

d. Calculate the surface area of the new prism. How many times greater is the surface area of the new prism?

What Is Your Answer?

3. **IN YOUR OWN WORDS** How does changing a dimension of a rectangular prism affect its surface area?

4. Why is knowing the surface area of an object important in a career as an aeronautical engineer?

Name ______________________________ Date __________

8.1 Practice

For use after Lesson 8.1

Find the surface area of the prism.

1.

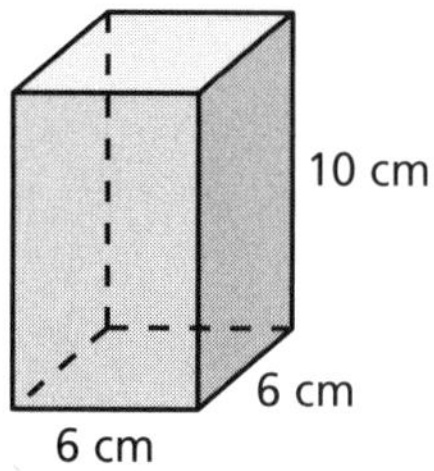

2.

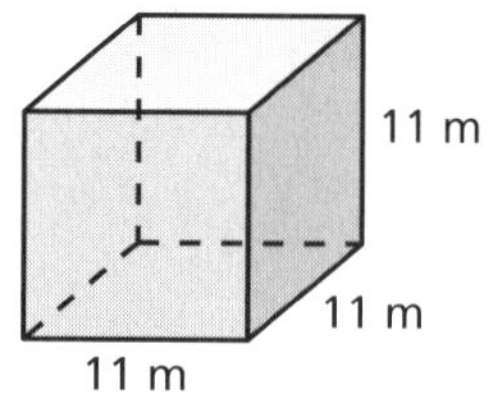

3.

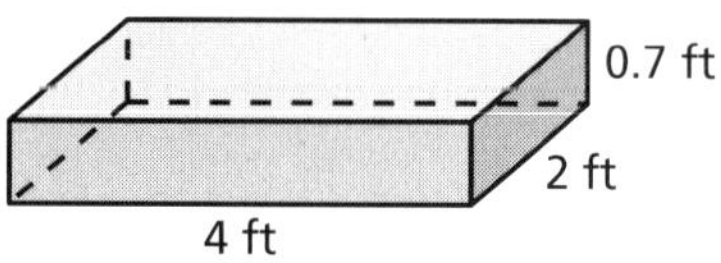

4.

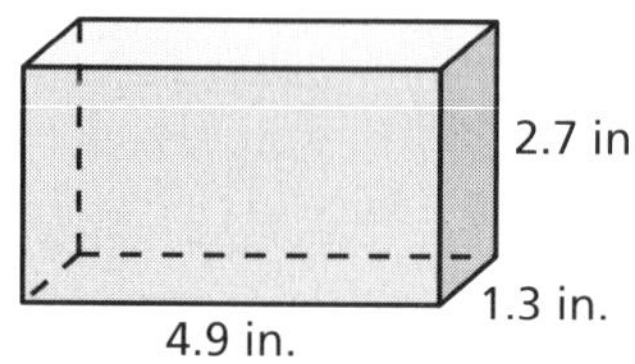

Compare the dimensions of the prisms. How many times greater is the surface area of the shaded prism than the surface area of the unshaded prism?

5.

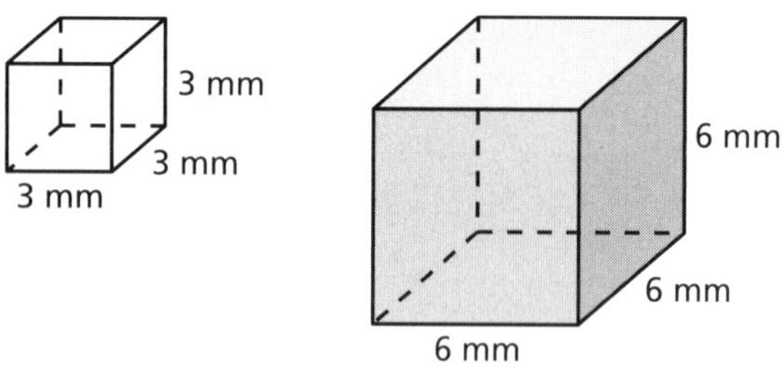

6.

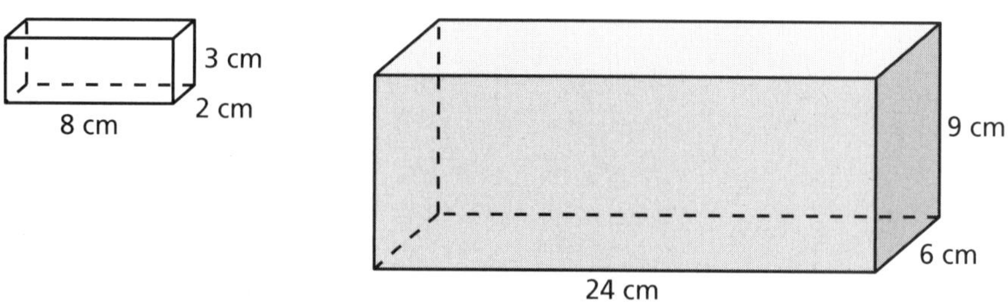

7. You are decorating two jewelry boxes with paper mache.

a. How much paper do you need to cover the larger box?

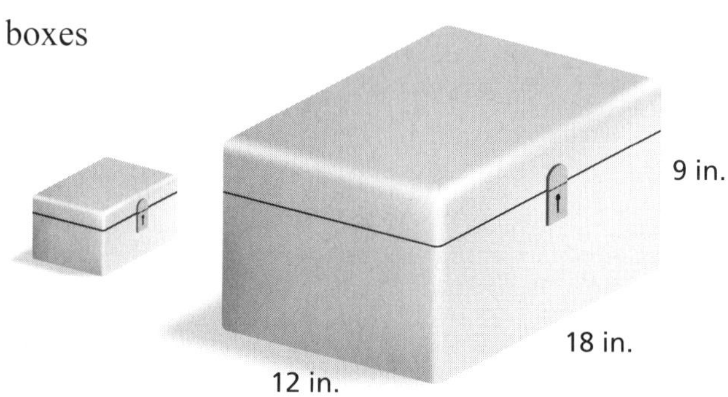

b. The side lengths of the smaller box are one-third the side lengths of the larger box. How much paper do you need to cover the smaller box?

Name____ Date__________

8.2 Volumes of Rectangular Prisms

For use with Activity 8.2

Essential Question How does changing a dimension of a rectangular prism affect its volume?

1 ACTIVITY: Comparing Volumes

Work with a partner.

a. Copy the net shown below on grid paper.

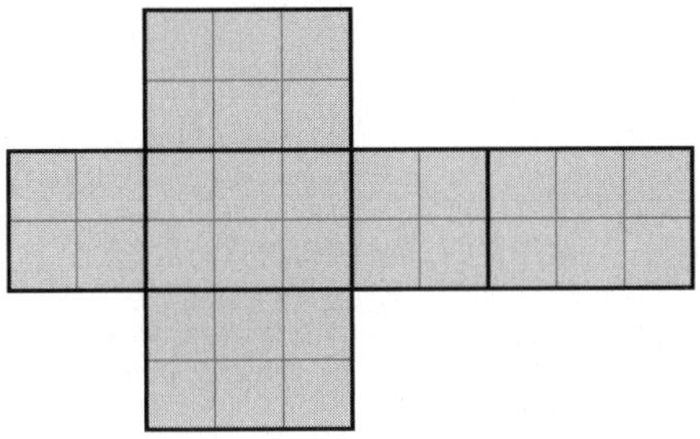

b. Double the width of the rectangular prism whose net is shown. Draw a net for the new prism.

c. Cut out both nets and fold them to form two prisms. Compare the two prisms visually. Estimate how much more volume the larger prism has than the smaller prism.

d. Find the actual volume of each prism. How close was your estimate?

Note: You used these nets in Activities 1 and 2 in Section 8.1.

Name ____ Date __________

8.2 Volumes of Rectangular Prisms (continued)

2 ACTIVITY: Comparing Volumes

Work with a partner.

a. Double all three dimensions of the original prism in Activity 1: the width, length, and height.

New width: ________________

New length: ________________

New height: ________________

b. Calculate the volume of the new prism. How many times greater is the volume of the new prism?

c. Triple all three dimensions of the original prism in Activity 1.

New width: ________________

New length: ________________

New height: ________________

d. Calculate the volume of the new prism. How many times greater is the volume of the new prism?

8.2 Volumes of Rectangular Prisms (continued)

What Is Your Answer?

3. **IN YOUR OWN WORDS** How does changing a dimension of a rectangular prism affect its volume?

4. **BIG IDEAS SCIENCE** When designing an airplane, which of the following best describes your goal? Explain your reasoning.

   a. Minimize surface area and maximize volume.

   b. Minimize volume and maximize surface area.

Name ________________________________ Date __________

8.2 Practice
For use after Lesson 8.2

Find the volume of the prism.

1.

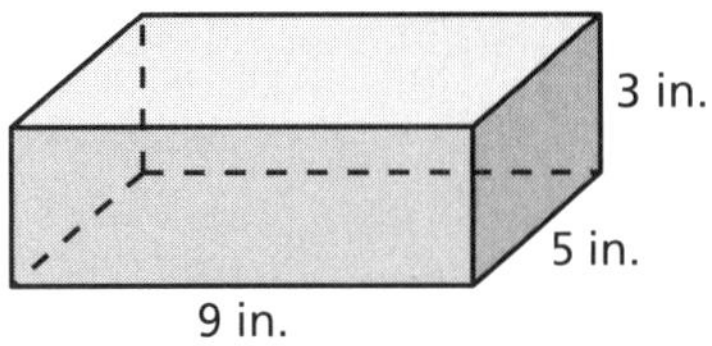

2.

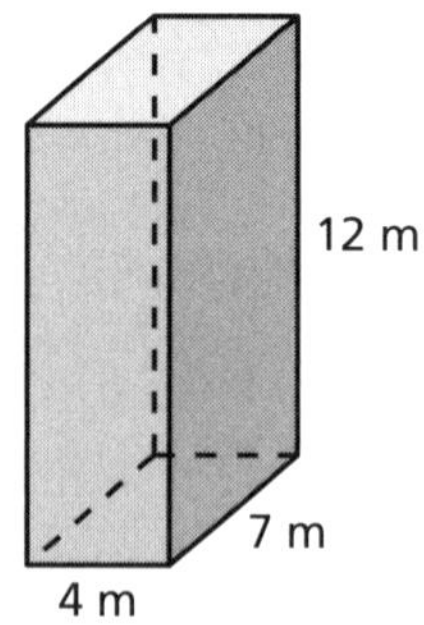

3.

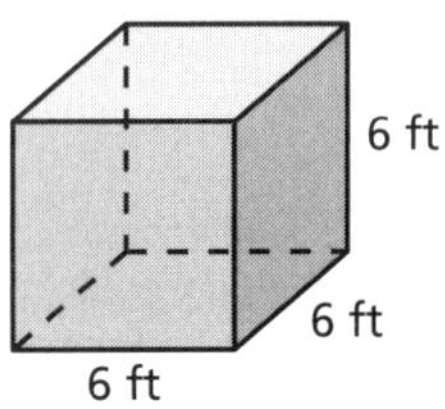

4.

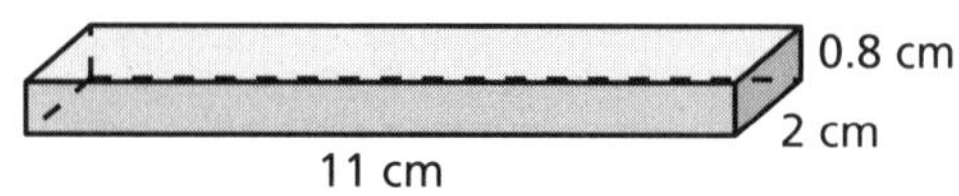

In Exercises 5 and 6, use the prism shown.

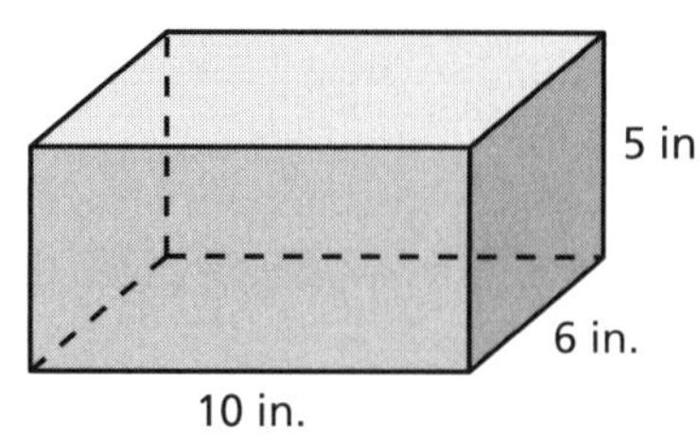

5. Double the width of the prism. How many times greater is the volume of the new prism than the volume of the original prism?

6. Triple the length of the prism. How many times greater is the volume of the new prism than the volume of the original prism.

7. A cooler is shown.

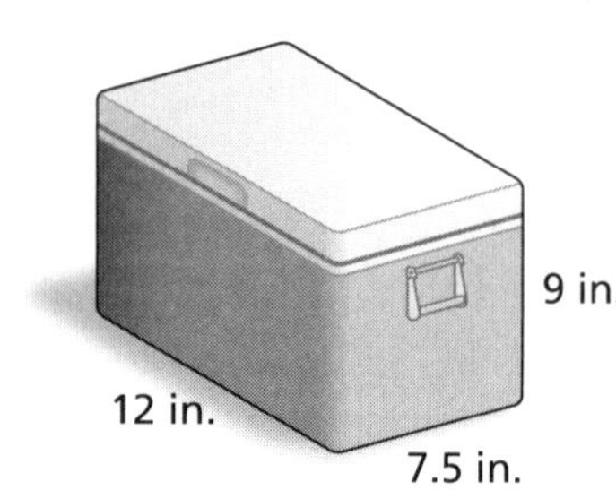

a. Find the volume of the cooler.

b. The dimensions of a freezer chest are four times greater than the dimensions of the cooler. What is the volume of the freezer chest (in cubic feet)?

Name ______________________________ Date __________

8.3 Surface Areas of Cylinders

For use with Activity 8.3

Essential Question How can you find the surface area of a cylinder?

A **cylinder** is a solid that has two parallel, congruent circular bases. The surface area of a cylinder is the sum of the areas of the bases and the lateral surface.

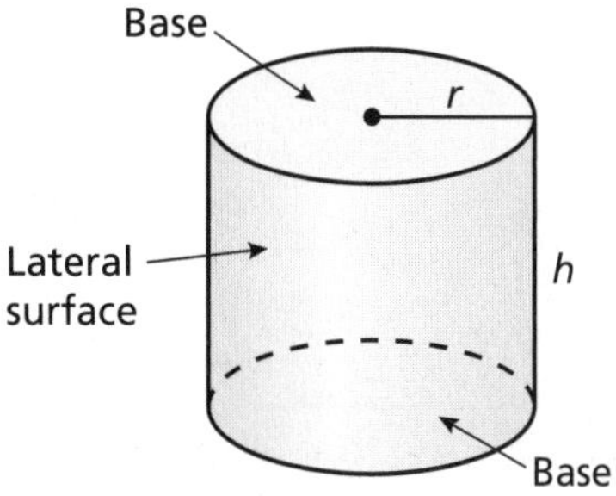

1 ACTIVITY: Finding Area

Work with a partner. Use a cardboard cylinder.

- **Talk about how you can find the area of the outside of the roll.**
- **Use a ruler to estimate the area of the outside of the roll.**
- **Cut the roll and press it out flat. Then find the area of the flattened cardboard. How close is your estimate to the actual area?**

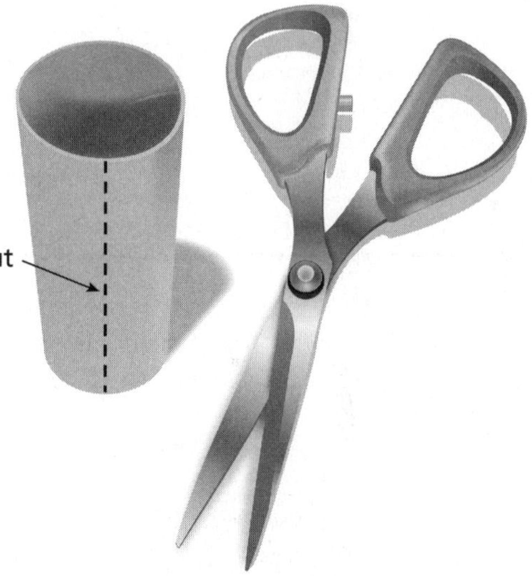

2 ACTIVITY: Finding Surface Area

Work with a partner.

- **Trace the top and bottom of a can on paper. Cut out the two shapes.**
- **Cut out a long paper rectangle. Make the width the same as the height of the can. Wrap the rectangle around the can. Cut off the excess paper so the edges just meet.**
- **Make a net for the can. Name the shapes in the net.**

Name ______________________ Date __________

8.3 Surface Areas of Cylinders (continued)

- **How are the dimensions of the rectangle related to the dimensions of the can?**

- **Explain how to use the net to find the surface area of the can.**

3 ACTIVITY: Estimation

Work with a partner. From memory, estimate the dimensions of the real-life items in parts (a)–(d) in inches. Then use the dimensions to estimate the surface area of each item in square inches.

a. 

b. 

Name________________________________ Date__________

c.

d.

What Is Your Answer?

4. **IN YOUR OWN WORDS** How can you find the surface area of a cylinder? Give an example with your description. Include a drawing of the cylinder.

5. To eight decimal places, $\pi \approx 3.14159265$. Which of the following is closest to π?

a. 3.14 **b.** $\frac{22}{7}$ **c.** $\frac{355}{113}$

"To approximate the irrational number $\pi \approx 3.141593$, I simply remember 1, 1, 3, 3, 5, 5."

"Then I compute the rational number $\frac{355}{113} \approx 3.141593$."

Name ______________________________ Date __________

8.3 Practice

For use after Lesson 8.3

Find the surface area of the cylinder. Round your answer to the nearest tenth.

1.

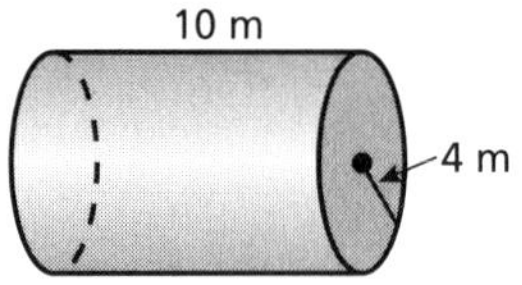

2.

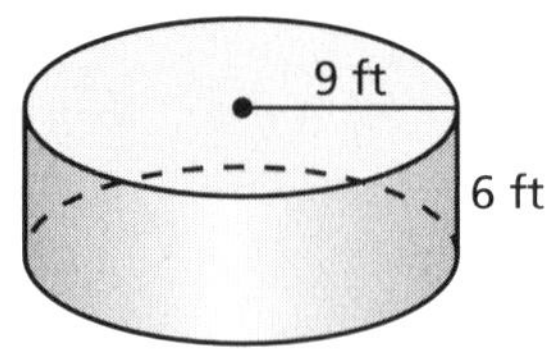

Find the lateral surface area of the cylinder. Round your answer to the nearest tenth.

3.

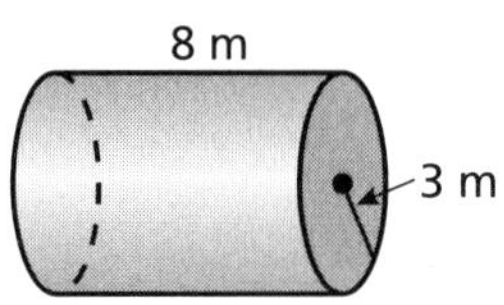

4.

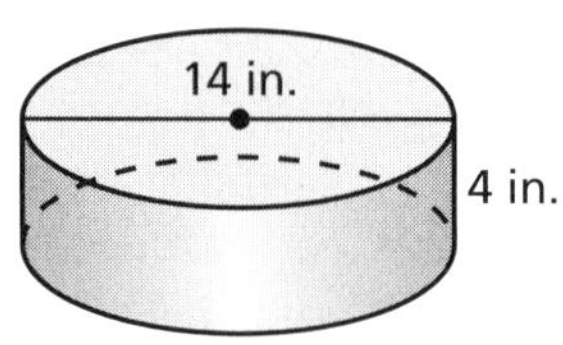

5. How much paper is used in the label for the can of cat food? Round your answer to the nearest whole number.

6. The circumference of a base of a cylinder is 33.3 inches and the height of the cylinder is 7.7 inches. What is the surface area of the cylinder?

Name____ Date__________

8.4 Volumes of Cylinders

For use with Activity 8.4

Essential Question How can you find the volume of a cylinder?

1 ACTIVITY: Finding a Formula Experimentally

Work with a partner.

a. Find the area of the face of a coin.

b. Find the volume of a stack of a dozen coins.

c. Generalize your results to find the volume of a cylinder.

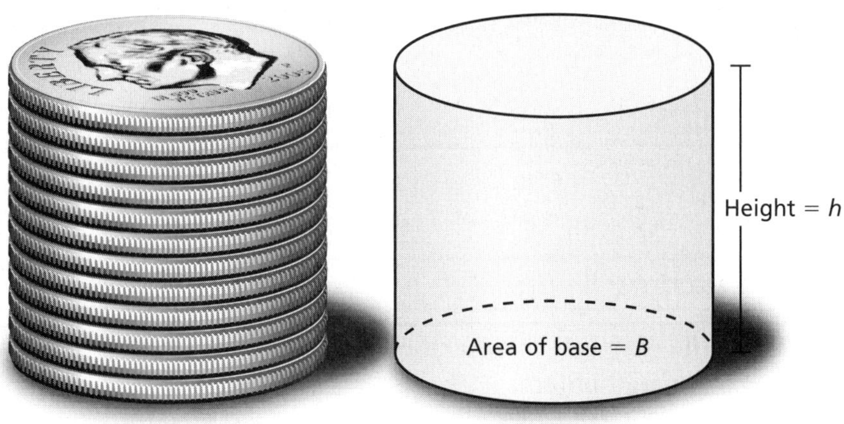

Name ____ Date __________

8.4 Volumes of Cylinders (continued)

2 ACTIVITY: Making a Business Plan

Work with a partner. You are planning to make and sell 3 different sizes of cylindrical candles. You buy 1 cubic foot of candle wax for \$20 to make 8 candles of each size.

a. Design the candles. What are the dimensions of each size?

b. You want to make a profit of \$100. Decide on a price for each size.

c. Did you set the prices so that they are proportional to the volume of each size of candle? Why or why not?

3 ACTIVITY: Science Experiment

Work with a partner. Use the diagram to describe how you can find the volume of a small object.

Name____ Date__________

8.4 Volumes of Cylinders (continued)

4 ACTIVITY: Comparing Cylinders

Work with a partner.

a. Just by looking at the two cylinders, which one do you think has the greater volume? Explain your reasoning.

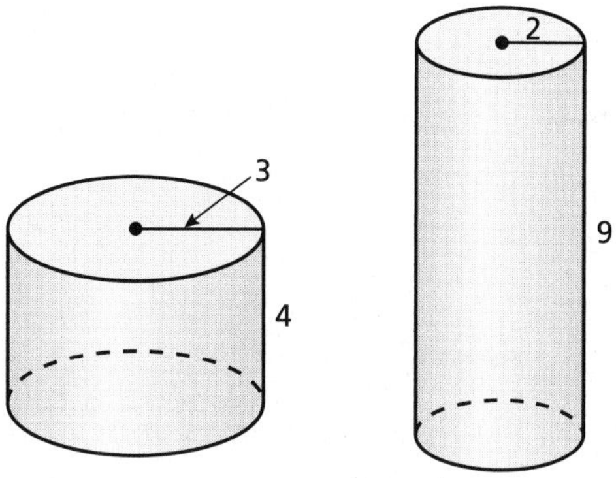

b. Find the volume of each cylinder. Was your prediction in part (a) correct? Explain your reasoning.

What Is Your Answer?

5. **IN YOUR OWN WORDS** How can you find the volume of a cylinder?

6. Compare your formula for the volume of a cylinder with the formula for the volume of a prism. How are they the same?

Name ____________________ Date ________

8.4 Practice
For use after Lesson 8.4

Find the volume of the cylinder. Round your answer to the nearest tenth.

1.
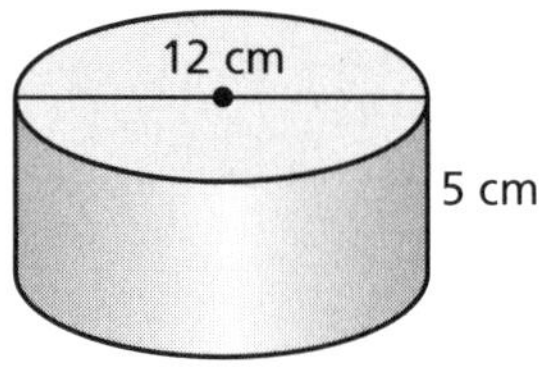

2.
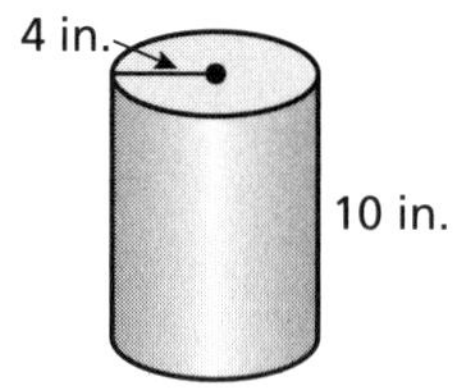

Find the height of the cylinder. Round your answer to the nearest whole number.

3. Volume = 84 $in.^3$

4. Volume = 650 cm^3
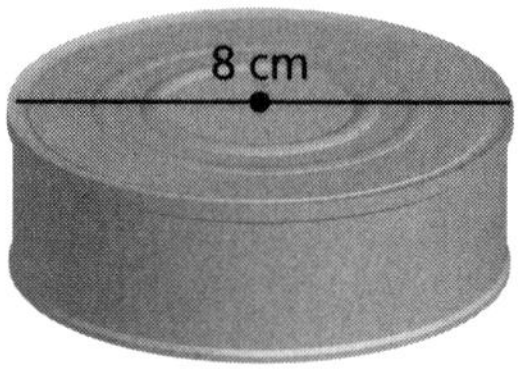

5. What happens to the volume of a cylinder if you double the radius?

6. To make orange juice, the directions call for a can of orange juice concentrate to be mixed with three cans of water. What is the volume of orange juice that you make?

Name____ Date__________

Chapter 9 Fair Game Review

Find the sum or difference.

1. 8.01 + 4.2

2. 9.736 + 10.922

3. 4.81 + 0.755

4. 10.6 − 2.18

5. 6.75 − 5.9

6. 3.874 − 0.06

7. You bring $20 on a shopping trip. You buy a shirt for $7.62.

a. How much money do you have after you buy the shirt?

b. Later, you buy a hat for $5.18. How much money do you have after you buy the hat?

8. You have 3.85 yards of fleece and 2.6 yards of silk for your design supplies. How many total yards of fabric do you have?

Name ______________________________ Date ________

Chapter 9 Fair Game Review (continued)

Find the product or quotient.

9. $3.92 \bullet 0.6$

10. $0.78 \bullet 0.13$

11. $\begin{array}{r} 5.004 \\ \times \quad 1.2 \\ \hline \end{array}$

12. $6.3 \div 0.7$

13. $2.25 \div 1.5$

14. $0.003\overline{)8.1}$

15. Grapes cost \$1.98 per pound. You buy 3.5 pounds of grapes. How much do you pay for the grapes?

16. A box of cereal costs \$3.69. The box has 13.7 ounces of cereal. About how much does each ounce cost?

Name____ Date __________

9.1 Exponents
For use with Activity 9.1

Essential Question How can you use exponents to write numbers?

The expression 3^5 is called a **power.** The base is 3. The **exponent** is 5.

base → 3^5 ← exponent

1 ACTIVITY: Using Exponent Notation

Work with a partner.

a. Complete the table.

Power	Repeated Multiplication Form	Value
$(-3)^1$		
$(-3)^2$		
$(-3)^3$		
$(-3)^4$		
$(-3)^5$		
$(-3)^6$		
$(-3)^7$		

b. Describe what is meant by the expression $(-3)^n$. How can you find the value of $(-3)^n$?

Name ____________________ Date ________

9.1 Exponents (continued)

2 ACTIVITY: Using Exponent Notation

Work with a partner.

a. The cube at the right has $3 in each of its small cubes. Write a single power that represents the total amount of money in the large cube.

b. Evaluate the power to find the total amount of money in the large cube.

3 ACTIVITY: Writing Powers as Whole Numbers

Work with a partner. Write each distance as a whole number. Which numbers do you know how to write in words? For instance, in words, 10^3 is equal to *one thousand*.

a. 10^{26} meters: Diameter of observable universe

b. 10^{21} meters: Diameter of Milky Way Galaxy

c. 10^{16} meters: Diameter of Solar System

d. 10^{7} meters: Diameter of Earth

e. 10^{6} meters: Length of Lake Erie shoreline

f. 10^{5} meters: Width of Lake Erie

Name____ Date__________

4 ACTIVITY: Writing a Power

Work with a partner. Write the number of kits, cats, sacks, and wives as a power.

As I was going to St. Ives
I met a man with seven wives
and every wife had seven sacks
and every sack had seven cats
and every cat had seven kits
Kits, cats, sacks, wives
How many were going to St. Ives?

Nursery Rhyme, 1730

What Is Your Answer?

5. **IN YOUR OWN WORDS** How can you use exponents to write numbers? Give some examples of how exponents are used in real life.

Name ____________________ Date ________

9.1 Practice

For use after Lesson 9.1

Write the product using exponents.

1. $4 \bullet 4 \bullet 4 \bullet 4 \bullet 4$

2. $\left(-\frac{1}{8}\right) \bullet \left(-\frac{1}{8}\right) \bullet \left(-\frac{1}{8}\right)$

3. $5 \bullet 5 \bullet (-x) \bullet (-x) \bullet (-x) \bullet (-x)$

4. $9 \bullet 9 \bullet y \bullet y \bullet y \bullet y \bullet y \bullet y$

Evaluate the expression.

5. 10^3

6. $(-7)^4$

7. $-\left(\frac{1}{6}\right)^5$

8. $3 + 6 \bullet (-5)^2$

9. $\left|-\frac{1}{3}\left(1^{10} + 9 - 2^3\right)\right|$

10. A foam toy is 2 inches wide. It doubles in size for every minute it is in water. Write an expression for the width of the toy after 5 minutes. What is the width after 5 minutes?

Name____ Date__________

9.2 Zero and Negative Exponents

For use with Activity 9.2

Essential Question How can you use zero and negative exponents in real life?

1 ACTIVITY: Investigating Powers of 10

In 1977, Charles and Ray Eames made the film *Powers of Ten* to illustrate the relative sizes of things in the Universe. Use the Internet to find and watch the film *Powers of Ten*.

Work with a partner. Some of the numbers given in the film are listed below. Write the exponent for each power of 10.

1 meter $= 10^0$ meter

10 meters $= 10^1$ meters

100 meters $= 10^2$ meters

1000 meters $= 10^3$ meters

10,000 meters $= 10^4$ meters

100,000 meters $= 10^5$ meters

1,000,000 microns $= 10^0$ meter

a. 100,000 microns $= 10^{\square}$ meter

b. 10,000 microns $= 10^{\square}$ meter

c. 1000 microns $= 10^{\square}$ meter

d. 100 microns $= 10^{\square}$ meter

e. 10 microns $= 10^{\square}$ meter

9.2 Zero and Negative Exponents (continued)

2 ACTIVITY: Writing a Report

Work with a partner. Use the Internet to research Charles and Ray Eames and how they made the film *Powers of Ten*. Then write a report that includes the following information.

- **Explain how the makers of the film knew what the far reaches of space and the inside of an atom looked like.**
- **Are all of the pictures actual photographs? Explain your reasoning.**
- **Describe several careers that involve outer space and atoms.**

3 ACTIVITY: Powers of 10 Matching Game

Work with a partner. Match each picture with its power of 10. Explain your reasoning.

10^5 m $\quad 10^2$ m $\quad 10^0$ m $\quad 10^{-1}$ m $\quad 10^{-2}$ m $\quad 10^{-5}$ m

A. 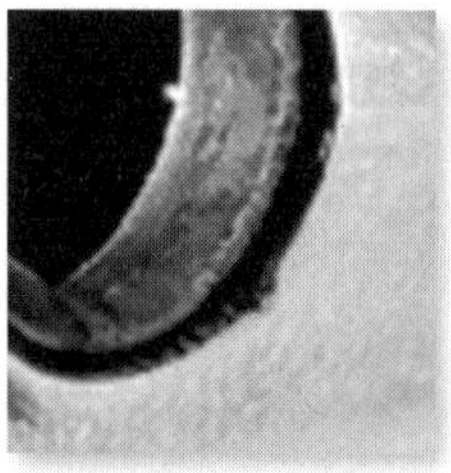

B. 

C. 

D. 

E. 

F. 

Name____ Date__________

9.2 Zero and Negative Exponents (continued)

4 ACTIVITY: Making a Powers of 10 Matching Game

Create your own matching game. Use the Internet to find at least 3 pictures that can be represented by different powers of 10. Trade games with a partner and play your partner's matching game.

What Is Your Answer?

5. **IN YOUR OWN WORDS** How can you use zero and negative exponents in real life? Give an example.

Name ____________________ Date ________

9.2 Practice

For use after Lesson 9.2

Evaluate the expression.

1. 29^0

2. 12^{-1}

3. $(-15)^{-2} \bullet (-15)^2$

4. $10^{-4} \bullet 10^{-6}$

5. $\frac{1}{3^{-3}} \bullet \frac{1}{3^5}$

6. $\frac{(4.1)^8}{(4.1)^7} + 9^0$

Simplify. Write your answer using only positive exponents.

7. $19x^{-6}$

8. $\frac{14}{a^{-8}}$

9. $\frac{16y^4}{4y^{10}}$

10. $3t^6 \bullet 8t^{-6}$

11. $7k^{-2} \bullet 5m^0 \bullet k^9$

12. $\frac{12s^0 \bullet 4^{-2} \bullet r^{-3}}{r^5}$

13. The density of a proton is about $\frac{1.64 \times 10^{-24}}{3.7 \times 10^{-38}}$ grams per cubic centimeter. Simplify the expression.

Name________________________ Date__________

9.3 Reading Scientific Notation

For use with Activity 9.3

Essential Question How can you read numbers that are written in scientific notation?

1 ACTIVITY: Very Large Numbers

Work with a partner.

- Use a calculator. Experiment with multiplying large numbers until your calculator gives an answer that is *not* in standard form.
- When the calculator on the right was used to multiply 2 billion by 3 billion, it listed the result as

  6.0E+18.

- Multiply 2 billion by 3 billion by hand. Use the result to explain what 6.0E+18 means.

- Check your explanation using products of other large numbers.
- Why didn't the calculator show the answer in standard form?

- Experiment to find the maximum number of digits your calculator displays. For instance, if you multiply 1000 by 1000 and your calculator shows 1,000,000, then it can display 7 digits.

Name ______________________________ Date __________

9.3 Reading Scientific Notation (continued)

2 ACTIVITY: Very Small Numbers

Work with a partner.

- Use a calculator. Experiment with multiplying very small numbers until your calculator gives an answer that is *not* in standard form.
- When the calculator at the right was used to multiply 2 billionths by 3 billionths, it listed the result as

  6.0E−18.

- Multiply 2 billionths by 3 billionths by hand. Use the result to explain what 6.0E−18 means.
- Check your explanation using products of other very small numbers.

3 ACTIVITY: Reading Scientific Notation

Work with a partner.

Each description gives an example of a number written in scientific notation. Answer the question in the description. Write your answer in standard form.

a. Nearly 1.0×10^5 dust mites can live in 1 square yard of carpet.

How many dust mites can live in 100 square yards of carpet?

b. A micron is about 4.0×10^{-5} inch. The length of a dust mite is 250 microns.

How long is a dust mite in inches?

9.3 Reading Scientific Notation (continued)

c. About 1.0×10^{15} bacteria live in a human body.

How many bacteria are living in the humans in your classroom?

d. A micron is about 4.0×10^{-5} inch. The length of a bacterium is about 0.5 micron.

How many bacteria could lie end-to-end on your finger?

e. Earth has only about 1.5×10^{8} kilograms of gold. Earth has a mass of 6.0×10^{24} kilograms.

What percent of Earth's mass is gold?

f. A gram is about 0.035 ounce. An atom of gold weighs about 3.3×10^{-22} gram.

How many atoms are in an ounce of gold?

What Is Your Answer?

4. **IN YOUR OWN WORDS** How can you read numbers that are written in scientific notation? Why do you think this type of notation is called "scientific notation"? Why is scientific notation important?

Name ______________________________ Date __________

9.3 Practice

For use after Lesson 9.3

Tell whether the number is written in scientific notation. Explain.

1. 14×10^{8}

2. 2.6×10^{12}

3. 4.79×10^{-8}

4. 3.99×10^{16}

5. 0.15×10^{22}

6. 6×10^{3}

Write the number in standard form.

7. 4×10^{9}

8. 2×10^{-5}

9. 3.7×10^{6}

10. 4.12×10^{-3}

11. 7.62×10^{10}

12. 9.908×10^{-12}

13. Light travels at 3×10^{8} meters per second.

    a. Write the speed of light in standard form.

    b. How far has light traveled after 5 seconds?

Name______________________________ Date__________

9.4 Writing Scientific Notation

For use with Activity 9.4

Essential Question How can you write a number in scientific notation?

1 ACTIVITY: Finding pH Levels

Work with a partner. In chemistry, pH is a measure of the activity of dissolved hydrogen ions (H^+). Liquids with low pH values are called acids. Liquids with high pH values are called bases.

Find the pH of each liquid. Is the liquid a base, neutral, or an acid?

a. Lime juice: $[H^+] = 0.01$

b. Egg: $[H^+] = 0.00000001$

c. Distilled water: $[H^+] = 0.0000001$

d. Ammonia water: $[H^+] = 0.00000000001$

e. Tomato juice: $[H^+] = 0.0001$

f. Hydrochloric acid: $[H^+] = 1$

pH	$[H^+]$	
14	1×10^{-14}	
13	1×10^{-13}	
12	1×10^{-12}	
11	1×10^{-11}	Bases
10	1×10^{-10}	
9	1×10^{-9}	
8	1×10^{-8}	
7	1×10^{-7}	Neutral
6	1×10^{-6}	
5	1×10^{-5}	
4	1×10^{-4}	
3	1×10^{-3}	Acids
2	1×10^{-2}	
1	1×10^{-1}	
0	1×10^{0}	

Name ______________________________ Date __________

9.4 Writing Scientific Notation (continued)

2 ACTIVITY: Writing Scientific Notation

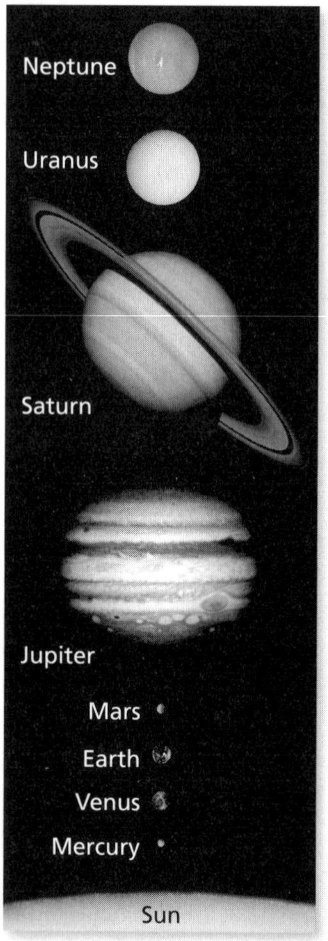

Work with a partner. Match each planet with its description. Then write each of the following in scientific notation.

- **Distance from the Sun (in miles)**
- **Distance from the Sun (in feet)**
- **Mass (in kilograms)**

a. **Distance:** 1,800,000,000 miles
Mass: 87,000,000,000,000,000,000,000,000 kg

b. **Distance:** 67,000,000 miles
Mass: 4,900,000,000,000,000,000,000,000 kg

c. **Distance:** 890,000,000 miles
Mass: 570,000,000,000,000,000,000,000,000 kg

d. **Distance:** 93,000,000 miles
Mass: 6,000,000,000,000,000,000,000,000 kg

e. **Distance:** 140,000,000 miles
Mass: 640,000,000,000,000,000,000,000 kg

f. **Distance:** 2,800,000,000 miles
Mass: 100,000,000,000,000,000,000,000,000 kg

g. **Distance:** 480,000,000 miles
Mass: 1,900,000,000,000,000,000,000,000,000 kg

h. **Distance:** 36,000,000 miles
Mass: 330,000,000,000,000,000,000,000 kg

Name____ Date__________

3 ACTIVITY: Making a Scale Drawing

Work with a partner. The illustration in Activity 2 is not drawn to scale. Make a scale drawing of the distances in our solar system.

- **Cut a sheet of paper into three strips of equal width. Tape the strips together.**
- **Draw a long number line. Label the number line in hundreds of millions of miles.**
- **Locate each planet's position on the number line.**

What Is Your Answer?

4. **IN YOUR OWN WORDS** How can you write a number in scientific notation?

Name ______________________________ Date __________

9.4 Practice

For use after Lesson 9.4

Write the number in scientific notation.

1. 4,200,000

2. 0.038

3. 600,000

4. 0.0000808

5. 0.0007

6. 29,010,000,000

Order the numbers from least to greatest.

7. $6.4 \times 10^8, 5.3 \times 10^9, 2.3 \times 10^8$

8. $9.1 \times 10^{-3}, 9.6 \times 10^{-3}, 9.02 \times 10^{-3}$

9. $7.3 \times 10^7, 5.6 \times 10^{10}, 3.7 \times 10^9$

10. $1.4 \times 10^{-5}, 2.01 \times 10^{-15}, 6.3 \times 10^{-2}$

11. A patient has 0.0000075 gram of iron in 1 liter of blood. The normal level is between 6×10^{-7} gram and 1.6×10^{-5} gram. Is the patient's iron level normal? Write the patient's amount of iron in scientific notation.

Name____ Date____________

9.5 Square Roots
For use with Activity 9.5

Essential Question How can you find the side length of a square when you are given the area of the square?

When you multiply a number by itself, you square the number.

Symbol for squaring is the exponent 2. → $4^2 = 4 \bullet 4$

$= 16$

4 squared is 16.

To "undo" this, take the **square root** of the number.

Symbol for square root is a radical sign. → $\sqrt{16} = \sqrt{4^2} = 4$ The square root of 16 is 4.

1 ACTIVITY: Finding Square Roots

Work with a partner. Use a square root symbol to write the side length of the square. Then find the square root. Check your answer by multiplying.

a. Sample: $s = \sqrt{121} =$ **Check:**

Area = 121 ft^2

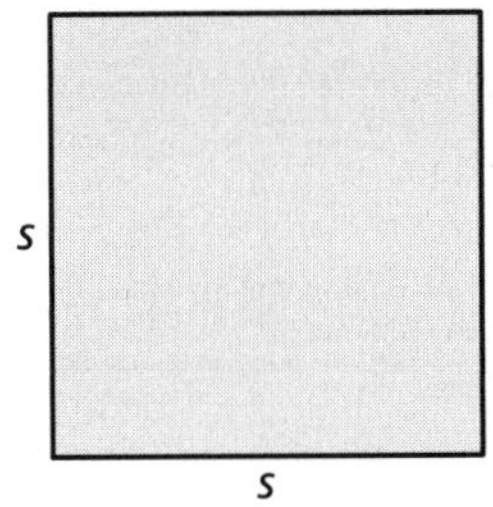

The length of each side of the square is ____________________.

b. Area = 81 yd^2

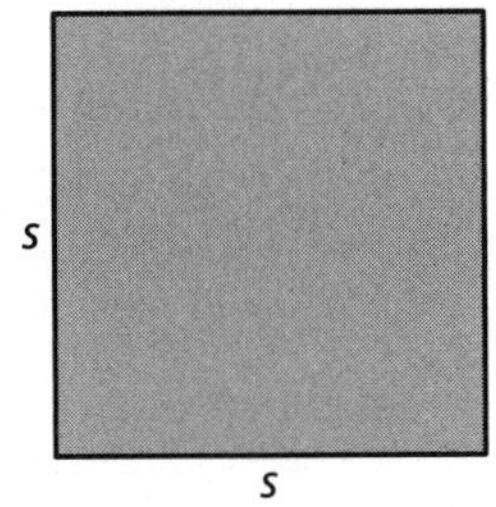

c. Area = 324 cm^2

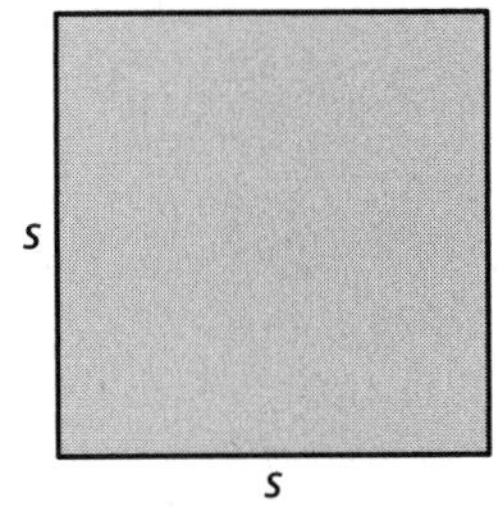

d. Area = 361 mi^2

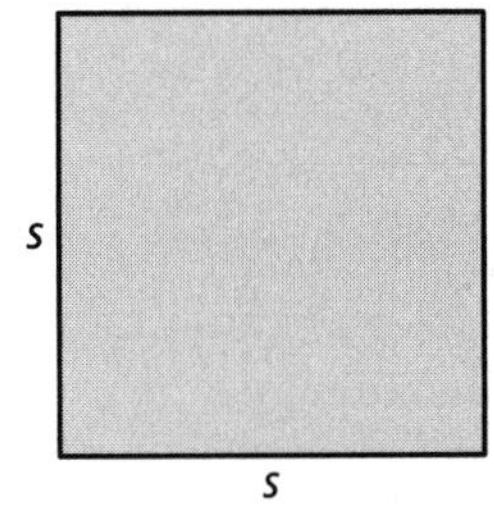

9.5 Square Roots (continued)

e. Area = 289 in.^{2}

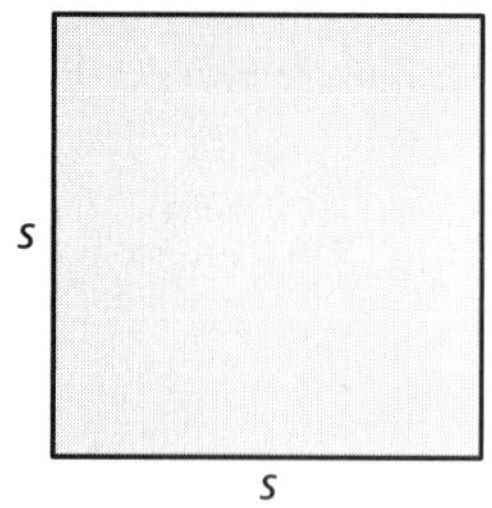

f. Area = 441 in.^{2}

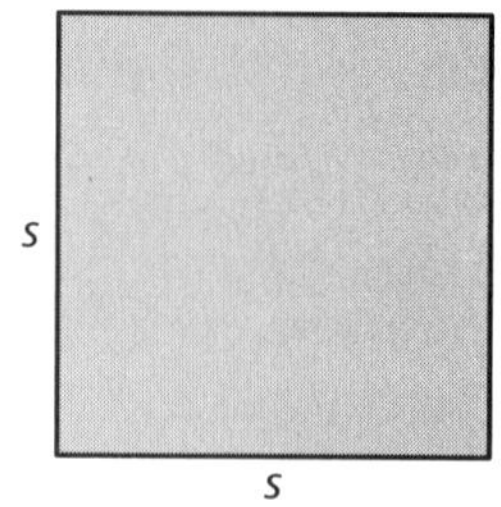

g. Area = 900 ft^2

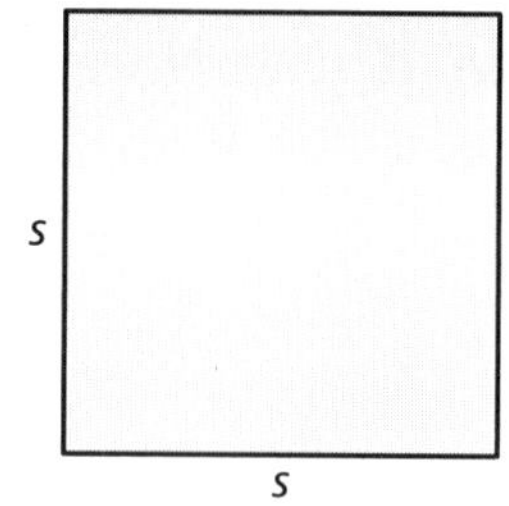

2 ACTIVITY: Describing Patterns

Work with a partner.

a. On a piece of grid paper, draw a sequence of small squares that represent the odd numbers.

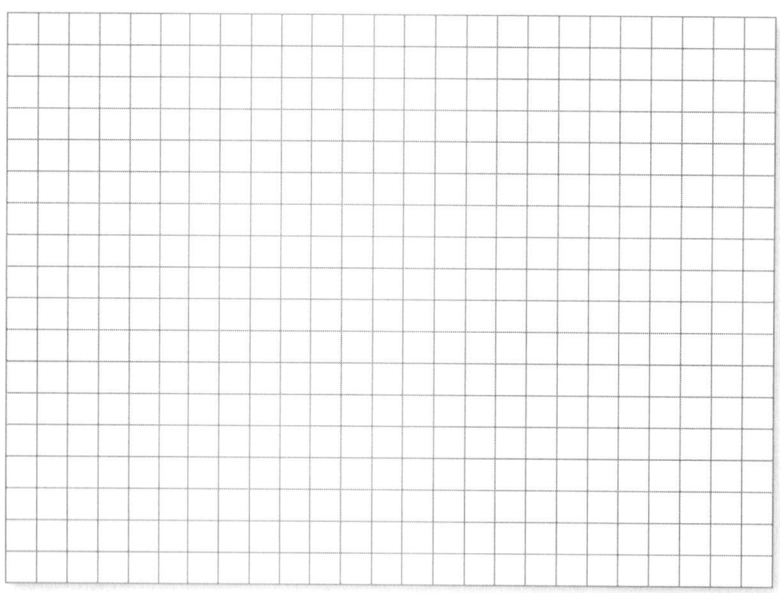

b. Draw a new sequence on the next page that represents the sums of terms in the sequence in part (a).

$$1, \qquad 1 + 3, \qquad 1 + 3 + 5, \ldots$$

c. Organize your results in a table.

9.5 Square Roots (continued)

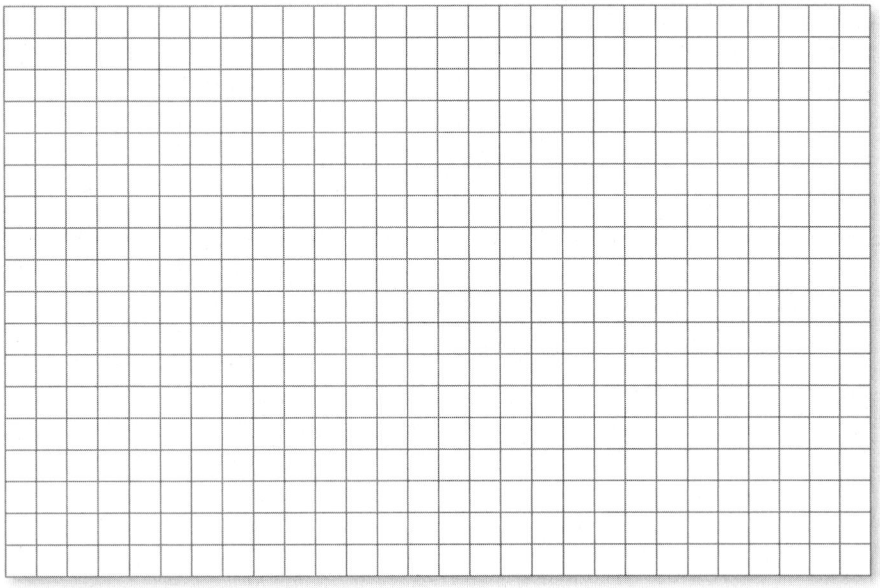

d. Describe the new sequence.

What Is Your Answer?

3. IN YOUR OWN WORDS How can you find the side length of a square when you are given the area of the square?

4. CONJECTURE Use the results of Activity 2 to write a conjecture about the sum of the first n odd numbers. On a different piece of grid paper, rearrange the small squares to explain *why* your conjecture is true.

Name ______________________________ Date __________

9.5 Practice

For use after Lesson 9.5

Find the two square roots of the number.

1. 9 **2.** 36 **3.** 144 **4.** 25

Find the square root(s).

5. $\sqrt{4}$ **6.** $-\sqrt{49}$ **7.** $\pm\sqrt{81}$

Evaluate the expression.

8. $27 + 11\sqrt{9}$ **9.** $2\sqrt{100} - 3\sqrt{16}$ **10.** $4\sqrt{81} - 5\sqrt{64}$

Find the radius of the circle. Use 3.14 for π.

11. Area $= 28.26 \text{ in.}^2$

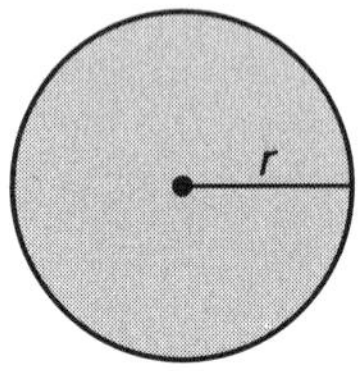

12. Area $= 379.94 \text{ cm}^2$

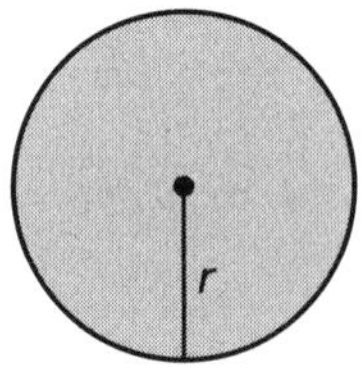

13. A mirror is shaped like a square. The area of the mirror is 64 square inches. What is the length of one side of the mirror?

14. The length of a rectangle is $\sqrt{36}$ meters. The width is $3\sqrt{25}$ meters. Write and evaluate an expression to find the perimeter of the rectangle.

Name____ Date__________

Chapter 10 Fair Game Review

The bar graph shows the favorite types of salad dressings of the students in a class.

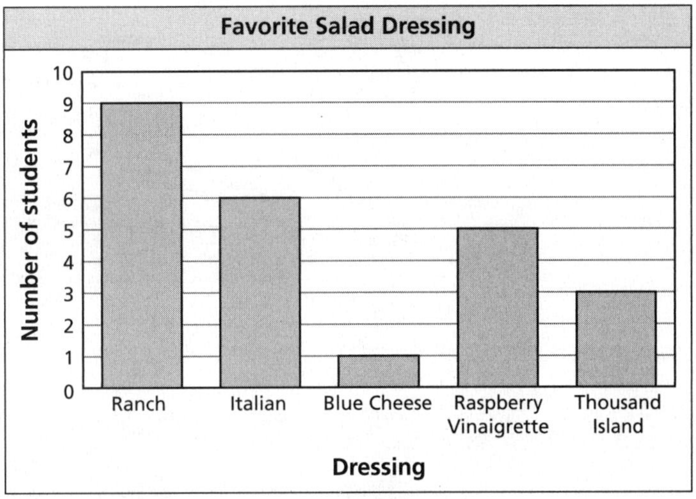

1. What salad dressing was chosen the most?

2. How many students said Raspberry Vinaigrette or Thousand Island is their favorite salad dressing?

3. How many students did *not* choose Italian as their favorite salad dressing?

4. How many students are in the class?

Name ______________________ Date __________

Chapter 10 Fair Game Review (continued)

Write the ratio in simplest form.

5. golf balls : baseballs

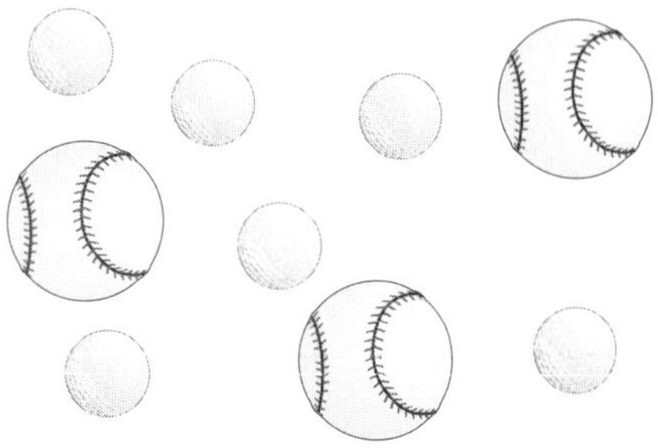

6. rulers to pencils

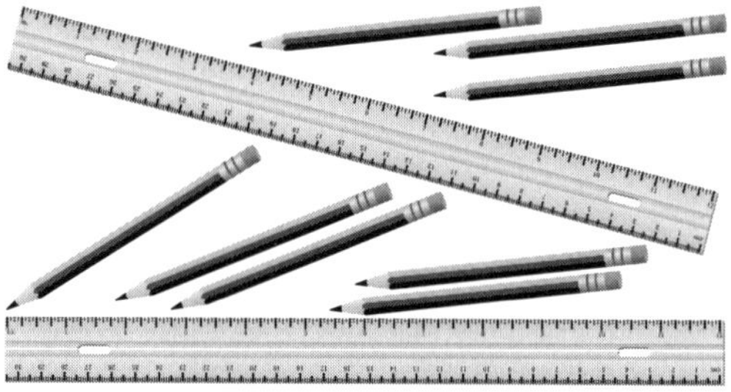

7. forks to total number of utensils

8. dog bone : toy mouse

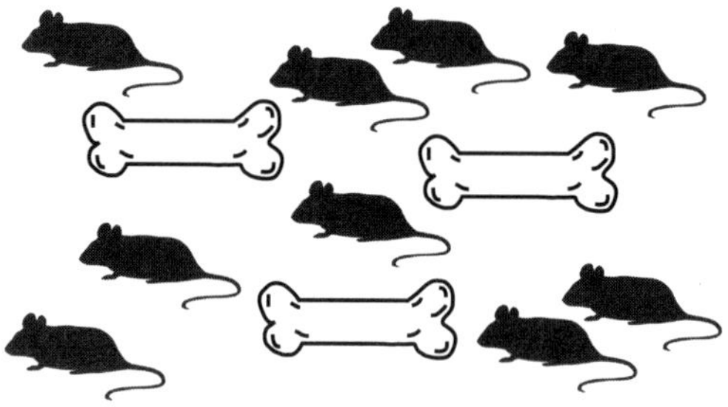

9. apples to strawberries

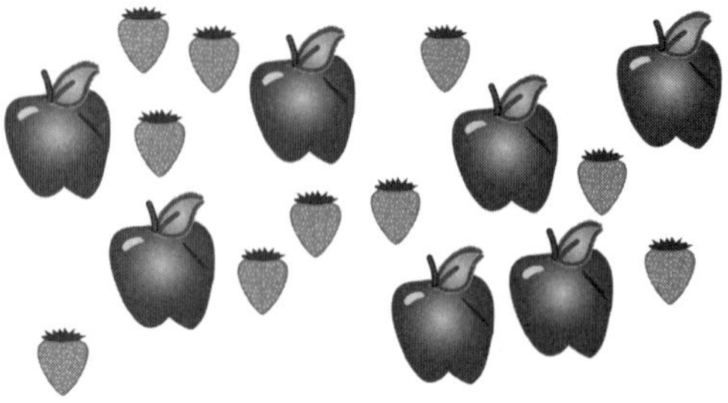

10. small fish : total number of fish

11. There are 24 flute players and 18 trumpet players in the band. Write the ratio of trumpet players to flute players.

Name______________________________ Date__________

10.1 Histograms
For use with Activity 10.1

Essential Question How can you use tables and graphs to help organize data?

1 ACTIVITY: Conducting an Experiment

Work with a partner.

a. Roll a number cube 20 times. Record your results in a tally chart.

	1	2	3	4	5	6
Tally						

Key: | = 1 𝍸 = 5

b. Make a bar graph of your totals.

c. Go to the board and enter your totals in the class tally chart.

d. Make a second bar graph showing the class totals. Compare and contrast the two bar graphs.

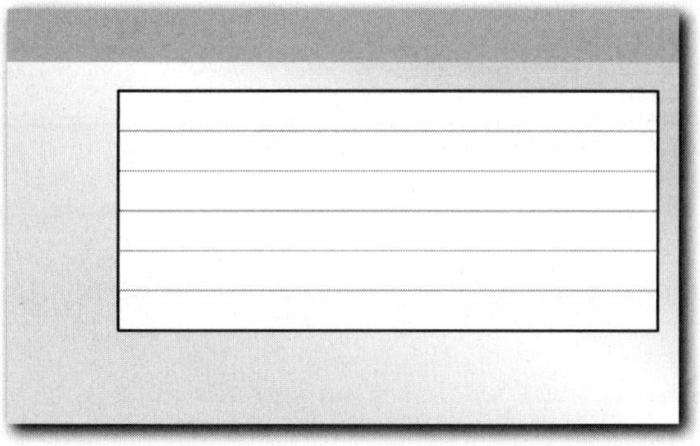

Name ______________________________ Date __________

10.1 Histograms (continued)

2 ACTIVITY: Organizing Data

Work with a partner. You are judging a paper airplane contest. Each contestant flies his or her paper airplane 20 times. Make a tally chart and a graph of the distances.

Complete the table and the graph using the data shown.

20.5 ft, 24.5 ft, 18.5 ft, 19.5 ft, 21.0 ft, 14.0 ft, 12.5 ft, 20.5 ft, 17.5 ft, 24.5 ft,
19.5 ft, 17.0 ft, 18.5 ft, 12.0 ft, 21.5 ft, 23.0 ft, 13.5 ft, 19.0 ft, 22.5 ft, 19.0 ft

Interval	Tally	Total
10.0–12.9		
13.0–15.9		
16.0–18.9		
19.0–21.9		
22.0–24.9		

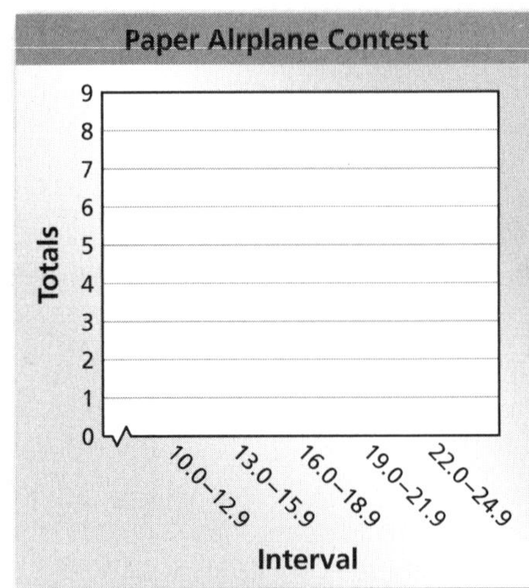

a. Make a different tally chart and graph of the distances using the following intervals.

10.0–11.9, 12.0–13.9, 14.0–15.9, 16.0–17.9,
18.0–19.9, 20.0–21.9, 22.0–23.9, 24.0–25.9

Name ______________________________ Date __________

10.1 Histograms (continued)

b. Which graph do you think represents the distances better? Explain.

3 ACTIVITY: Developing an Experiment

Work with a partner.

a. Design and make a paper airplane from a single sheet of $8\frac{1}{2}$-by-11 inch paper.

b. Fly the airplane 20 times. Keep track of the distance flown each time.

Flight	**1**	**2**	**3**	**4**	**5**	**6**	**7**	**8**	**9**	**10**
Distance										
Flight	**11**	**12**	**13**	**14**	**15**	**16**	**17**	**18**	**19**	**20**
Distance										

c. Organize your results in a tally chart and a graph. What is the mean distance flown by the airplane?

What Is Your Answer?

4. IN YOUR OWN WORDS How can you use tables and graphs to help organize data? Give examples of careers in which the organization of data is important.

Name ______________________________ Date __________

10.1 Practice

For use after Lesson 10.1

Display the data in a histogram.

1.

Birthdays	
Months	**Frequency**
Jan–Mar	15
Apr–June	9
Jul–Sept	6
Oct–Dec	12

2.

Goals Scored	
Goals	**Frequency**
0–2	6
3–5	8
6–8	2
9–11	1

3.

Height Jumped	
Inches	**Frequency**
0–11	7
12–23	10
24–35	5
36–47	2

4.

Money Spent	
Dollars	**Frequency**
0–19	3
20–39	8
40–59	8
60–79	15

5. The histogram shows the times students ran the mile in gym class.

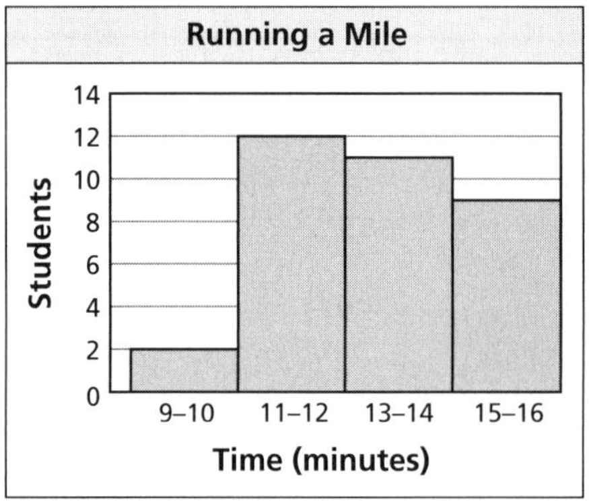

a. Which interval contains the fewest data values?

b. How many students are in the class?

c. What percent of students ran the mile in 12 minutes or less?

Name____ Date__________

10.2 Fundamental Counting Principle

For use with Activity 10.2

Essential Question How can you count the number of different combinations that are possible on a lock?

1 ACTIVITY: Comparing Combination Locks

Work with a partner. You are buying a combination lock for your locker. You have three choices. For which of the following locks is it most difficult to guess the combination? Explain your reasoning.

a. This lock has 3 wheels, each numbered from 0 to 9.

b. This lock is numbered from 0 to 39. Each combination uses three numbers in a right, left, right pattern.

c. This lock has 4 wheels:

Wheel 1: 0–9

Wheel 2: A–J

Wheel 3: K–T

Wheel 4: 0–9

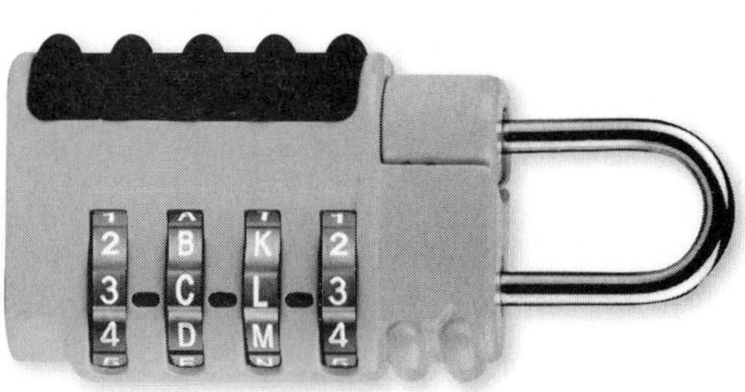

Name ______________________________ Date __________

10.2 Fundamental Counting Principle (continued)

2 ACTIVITY: Comparing Password Security

Work with a partner. Which of the following password requirements is most secure? Explain your reasoning. Include the number of different passwords that are possible for each requirement.

a. The password must have four digits.

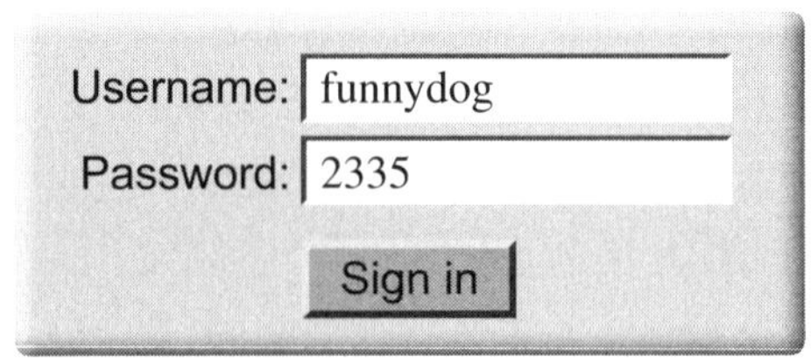

b. The password must have five digits.

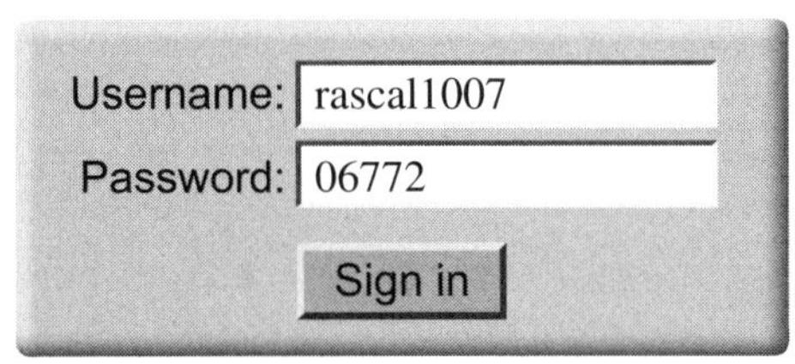

c. The password must have six letters.

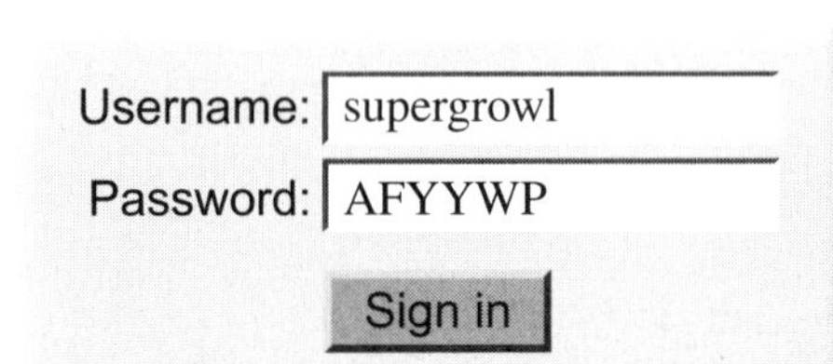

d. The password must have eight digits or letters.

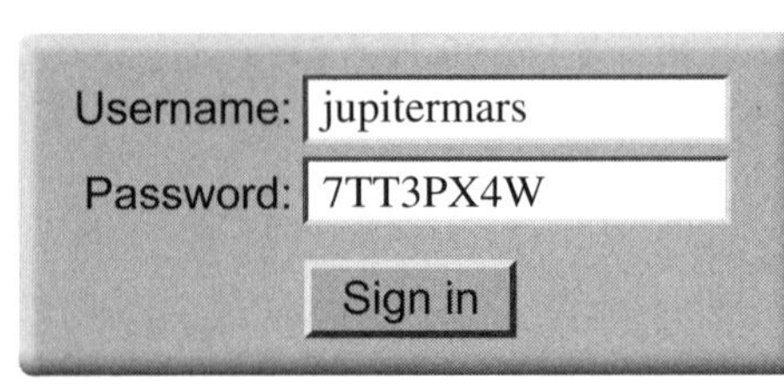

What Is Your Answer?

3. **IN YOUR OWN WORDS** How can you count the number of different combinations that are possible on a lock?

4. **SECURITY** A hacker uses a software program to guess the passwords in Activity 2. The program checks 600 passwords per minute. How long will it take the program to guess each of the four types of passwords?

   a. four digits

   b. five digits

   c. six letters

   d. eight digits or letters

Name ____________________ Date ________

10.2 Practice

For use after Lesson 10.2

1. Use a tree diagram to find the total number of possible outcomes.

Bed Sheets	
Size	Twin, Twin XL, Full, Queen, King
Style	Solid, Patterned

Use the Fundamental Counting Principle to find the total number of possible outcomes.

2.

Photos	
Size	Wallet, 4 by 6, 5 by 7, 8 by 10, 11 by 14, 16 by 20
Finish	Matte, Glossy
Edits	Red eye, Black and white, Crop

3.

Laptops	
Hard Drive	250 GB, 320 GB, 500 GB
Style	HD, LCD
Color	Black, White, Red, Blue, Pink, Green, Purple

4. Each question on a multiple choice quiz has 4 choices. There are 10 questions on the quiz. Use the Fundamental Counting Principle to find the total number of possible ways to answer the quiz.

5. You have 12 sports trophies that you want to display in your room.

a. Use the Fundamental Counting Principle to find the total number of ways you can arrange the trophies.

b. Why should you use the Fundamental Counting Principle instead of a tree diagram to find the total number of ways you can arrange the trophies?

Name____ Date__________

10.3 Theoretical Probability
For use with Activity 10.3

Essential Question How can you find a theoretical probability?

1 ACTIVITY: Black and White Spinner Game

Work with a partner. You work for a game company. You need to create a game that uses the spinner below.

a. Write rules for a game that uses the spinner. Then play it.

b. After playing the game, do you want to revise the rules? Explain.

c. Each pie-shaped section of the spinner is the same size. What is the measure of the central angle of each section?

d. What is the probability that the spinner will land on 1? Explain.

Name ______________________________ Date __________

10.3 Theoretical Probability (continued)

2 ACTIVITY: Changing the Spinner

Work with a partner. For each spinner, find the probability of landing on each number. Do your rules for Activity 1 make sense for these spinners? Explain.

a.

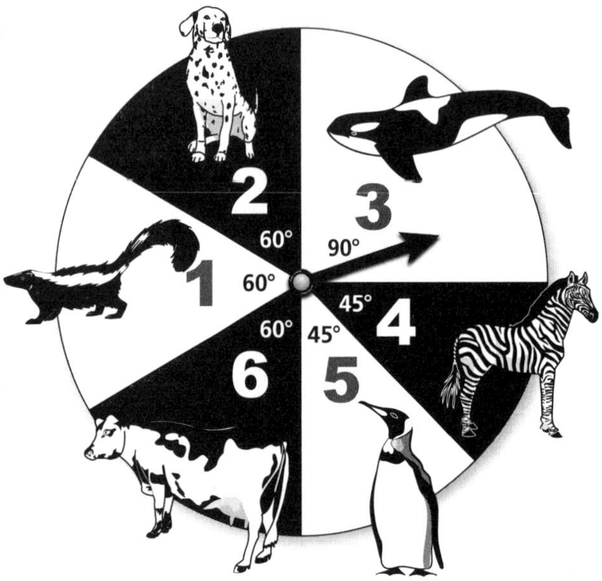

b.

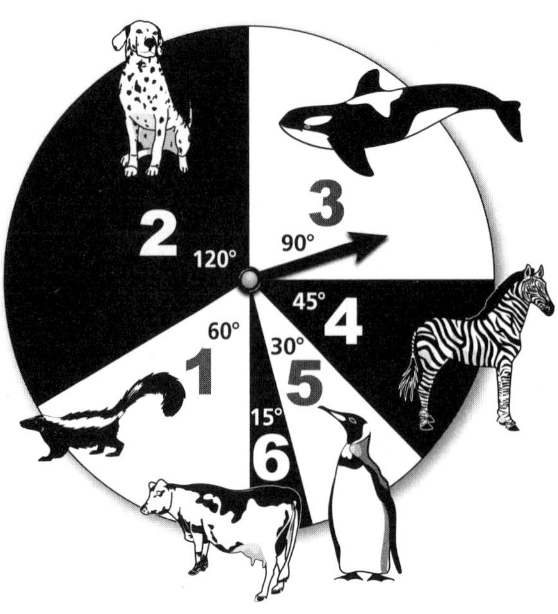

10.3 Theoretical Probability (continued)

3 ACTIVITY: Is This Game Fair?

Work with a partner. Apply the following rules to each spinner in Activities 1 and 2. Is the game fair? If not, who has the better chance of winning?

- **Take turns spinning the spinner.**
- **If the spinner lands on an odd number, Player 1 wins.**
- **If the spinner lands on an even number, Player 2 wins.**

What Is Your Answer?

4. **IN YOUR OWN WORDS** How can you find a theoretical probability?

5. Find and describe a career in which probability is used. Explain why probability is used in that career.

6. Two people play the following game.

Each player has 6 cards numbered 1, 2, 3, 4, 5, and 6. At the same time, each player holds up one card. If the product of the two numbers is odd, Player 1 wins. If the product is even, Player 2 wins. Continue until both players are out of cards. Which player is more likely to win? Why?

Name ______________________________ Date __________

10.3 Practice

For use after Lesson 10.3

Use a number cube to determine the theoretical probability of the event.

1. Rolling a 2

2. Rolling a 5

3. Rolling an even number

4. Rolling a number greater than 1

A spinner is used for a game. Determine if the game is fair. If it is *not* fair, who has the greater probability of winning?

5. You win if the number is less than 4. If it is not less than 4, your friend wins.

6. You win if the number is a multiple of 2. If it is not a multiple of 2, your friend wins.

7. At a carnival, you pick a duck out of a pond that designates a prize. You want to win a large prize and the theoretical probability of winning it is $\frac{9}{25}$. There are 50 ducks. How many ducks will win a large prize?

Name______________________________ Date__________

10.4 Experimental Probability

For use with Activity 10.4

Essential Question What is meant by experimental probability?

1 ACTIVITY: Throwing Sticks

Play with a partner. This game is based on an Apache game called "Throw Sticks."

- **Take turns throwing three sticks into the center of the circle and moving around the circle according to the chart.**
- **If your opponent lands on or passes your playing piece, you must start over.**
- **The first player to pass his or her starting point wins.**

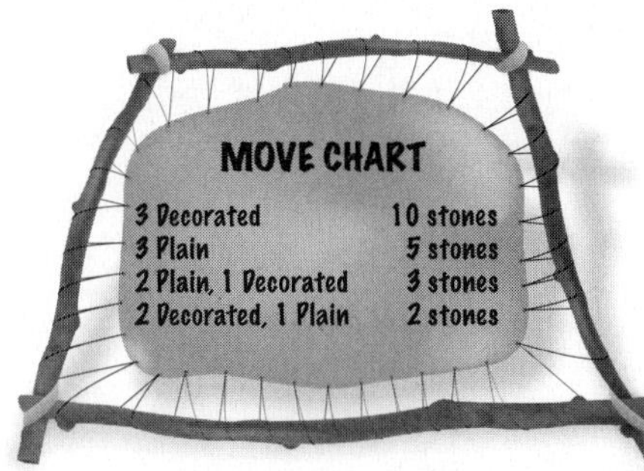

Each stick has one plain side and one decorated side.

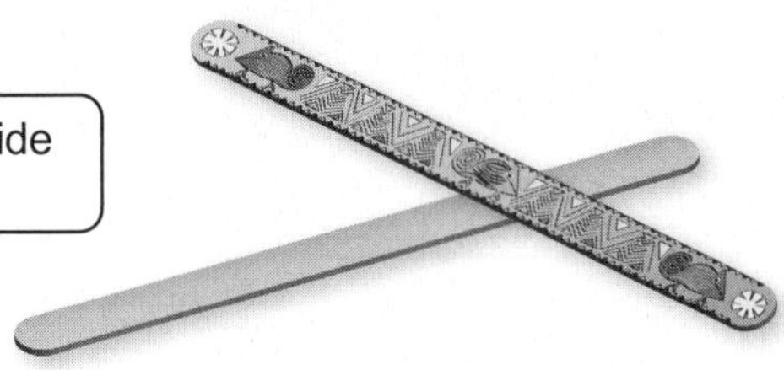

The game board has 40 stones arranged in a circle. The stones are placed in groups of 10.

Players start on opposite sides of the circle.

Player 1 Starting Point

Player 2 Starting Point

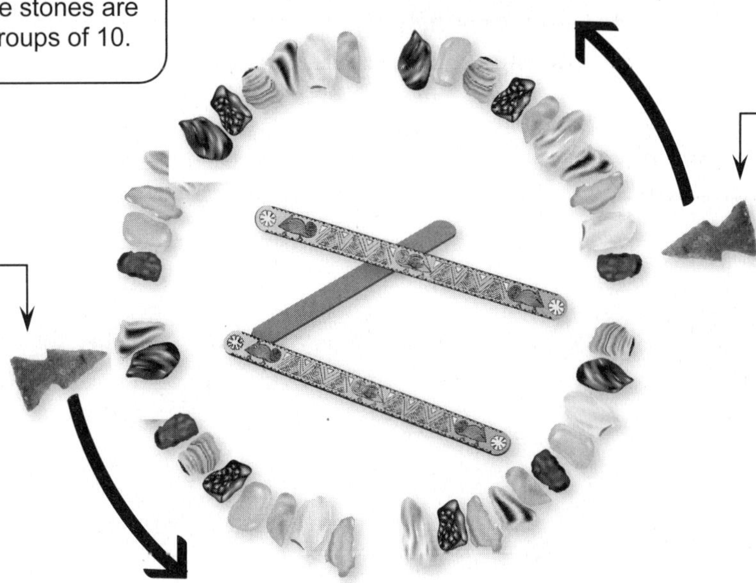

10.4 Experimental Probability (continued)

2 ACTIVITY: Conducting an Experiment

Work with a partner. Throw the 3 sticks 32 times. Tally the results using the outcomes listed below. A "P" represents the plain side landing up and a "D" represents the decorated side landing up. Organize the results in a bar graph. Use the bar graph to estimate the probability of each outcome. These are called experimental probabilities.

a. PPP

b. DPP

c. DDP

d. DDD

3 ACTIVITY: Analyzing the Possibilities

Work with a partner. A tree diagram helps you see different ways that the same outcome can occur.

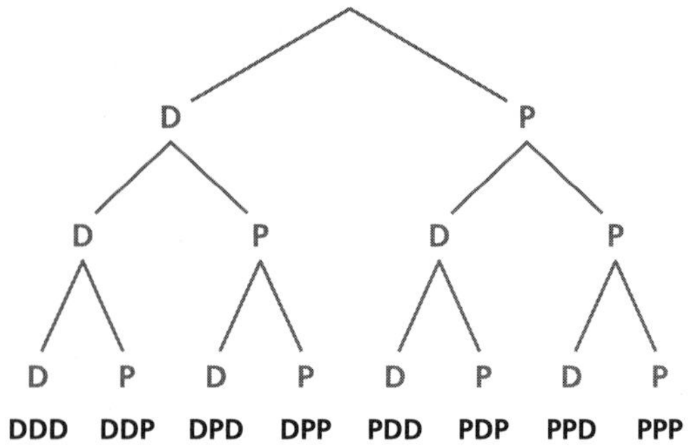

a. Find the number of ways that each outcome can occur.

- Three Ps
- One D and two Ps
- Two Ds and one P
- Three Ds

b. Find the theoretical probability of each outcome.

c. Compare and contrast your experimental and theoretical probabilities.

What Is Your Answer?

4. IN YOUR OWN WORDS What is meant by experimental probability?

5. Give a real-life example of experimental probability.

Name ______________________________ Date __________

10.4 Practice

For use after Lesson 10.4

Use the bar graph to find the experimental probability of the event.

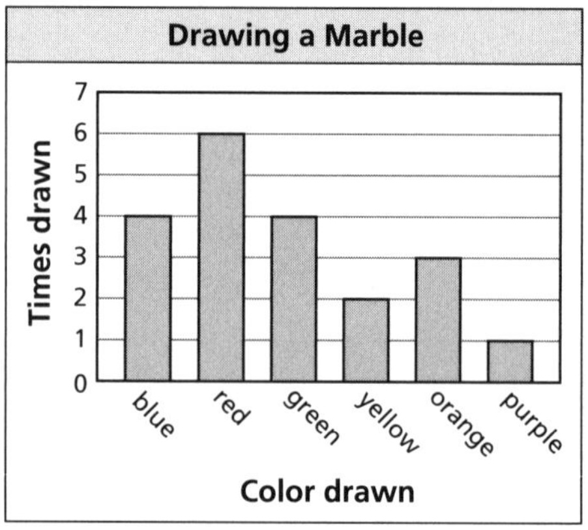

1. Drawing red

2. Drawing orange

3. Drawing *not* yellow

4. Drawing a color with more than 4 letters in its name

5. There are 25 students' names in a hat. You choose 5 names. Three are boys' names and two are girls' names. How many of the 25 names would you expect to be boys' names?

6. You must stop at 3 of 5 stoplights on a stretch of road. If this trend continues, how many times will you stop if the road has 10 stoplights?

7. Your teacher has a large box containing an equal number of red and blue folders. There are 24 students in your class. The teacher passes out the folders at random. Ten students receive a red folder. Compare the experimental probability of receiving a red folder with the theoretical probability of receiving a red folder.

Name ______________________________ Date __________

10.5 Compound Events

For use with Activity 10.5

Essential Question How can you find the probability that a compound event will occur?

1 ACTIVITY: Performing an Experiment

Work with a partner.

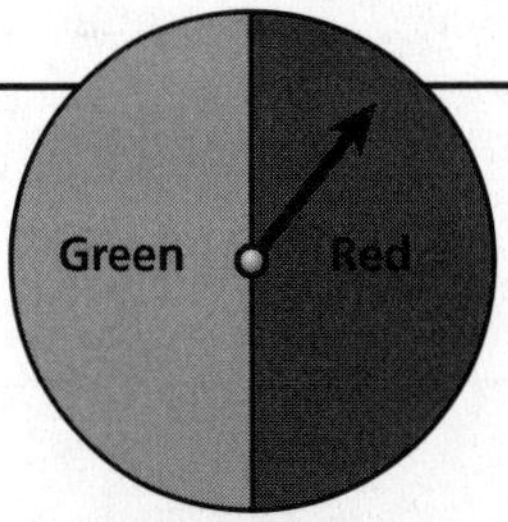

a. If you spin each spinner 48 times, how many times would you expect the two colors that the spinners land on to make purple? Explain your reasoning.

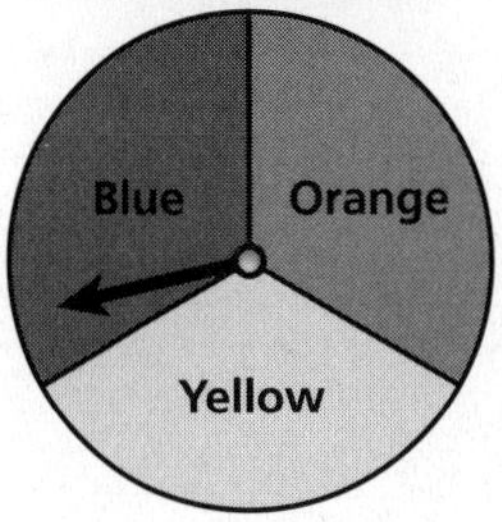

b. Spin each spinner 48 times and record your results in the table.* How many times did the colors make purple? Compare the result with your answer to part (a).

Trial	Purple	Not Purple
1		
2		
3		
4		
5		
6		
7		
8		
9		
10		
11		
12		
13		
14		
15		
16		

Trial	Purple	Not Purple
17		
18		
19		
20		
21		
22		
23		
24		
25		
26		
27		
28		
29		
30		
31		
32		

Trial	Purple	Not Purple
33		
34		
35		
36		
37		
38		
39		
40		
41		
42		
43		
44		
45		
46		
47		
48		

*Cut-outs are available in the back of the Record and Practice Journal.

2 ACTIVITY: Performing an Experiment

Work with a partner.

a. There are five pieces of paper in a bag: one red, one blue, one yellow, one orange, and one green. You randomly draw one piece of paper from the bag, keep it, and then draw a second piece of paper. If you do this 48 times, how many times would you expect the two colors chosen to make purple? Explain.

b. Perform the experiment 48 times. How many times did the colors make purple? Compare the result with your answer to part (a).

Trial	Purple	Not Purple
1		
2		
3		
4		
5		
6		
7		
8		
9		
10		
11		
12		
13		
14		
15		
16		

Trial	Purple	Not Purple
17		
18		
19		
20		
21		
22		
23		
24		
25		
26		
27		
28		
29		
30		
31		
32		

Trial	Purple	Not Purple
33		
34		
35		
36		
37		
38		
39		
40		
41		
42		
43		
44		
45		
46		
47		
48		

What Is Your Answer?

3. **IN YOUR OWN WORDS** How can you find the probability that a compound event will occur?

4. Compare the experiments and their results in Activities 1 and 2.

Name ______________________________ Date __________

10.5 Practice

For use after Lesson 10.5

You roll a number cube twice. Find the probability of the events.

1. Rolling a 3 twice

2. Rolling an even number and a 5

3. Rolling an odd number and a 2 or a 4

4. Rolling a number less than 6 and a 3 or a 1

You randomly choose a letter from a hat with the letters A through J. Without replacing the first letter, you choose a second letter. Find the probability of the events.

5. Choosing an H and then a D

6. Choosing a consonant and then an E or an I

7. Choosing a vowel and then an F

8. Choosing a vowel and then a consonant

9. You have 3 clasp bracelets, 4 watches, and 5 stretch bracelets. You randomly choose two from your jewelry box. What is the probability that you will choose 2 watches?

10. There is a pile of football cards and a pile of baseball cards at a garage sale. The probability that you randomly choose a quarterback from the football cards is $\frac{1}{15}$. The probability that you randomly choose a quarterback and a pitcher is $\frac{1}{315}$.

a. What is the probability that you choose a pitcher from the baseball cards?

b. There are 6 pitcher baseball cards. How many baseball cards are in the pile?

Glossary

This student friendly glossary is designed to be a reference for key vocabulary, properties, and mathematical terms. Several of the entries include a short example to aid your understanding of important concepts.

Also available at *BigIdeasMath.com*:

- multi-language glossary
- vocabulary flash cards

absolute value

The distance between a number and 0 on a number line; The absolute value of a number a is written as $|a|$.

$$|-5| = 5$$

$$|5| = 5$$

acute angle

An angle whose measure is less than $90°$

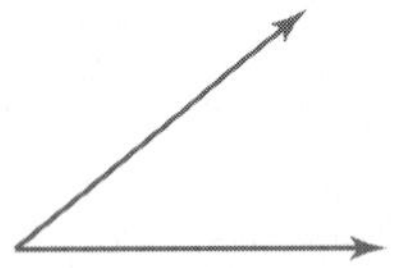

addend

A number to be added to another number

2 or 3 in the sum $2 + 3$

Addition Property of Equality

Adding the same number to each side of an equation produces an equivalent equation.

$$\begin{array}{rcr} x - 5 & = & -1 \\ \underline{+\,5} & & \underline{+\,5} \\ x & = & 4 \end{array}$$

Addition Property of Inequality

If you add the same number to each side of an inequality, the inequality remains true.

$$\begin{array}{rcr} x - 3 & > & -10 \\ \underline{+\,3} & & \underline{+\,3} \\ x & > & -7 \end{array}$$

Addition Property of Zero

The sum of any number and 0 is that number.

$$5 + 0 = 5$$

additive inverse

The opposite of a number

The additive inverse of 8 is -8.

Additive Inverse Property

The sum of an integer and its additive inverse is 0.

$$8 + (-8) = 0$$

algebraic expression

An expression that contains numbers, operations, and one or more variables

$$8 + x,\ 6 \times a - b$$

angle

A figure formed by two rays with the same endpoint

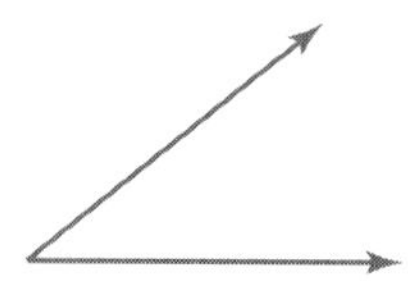

angle of rotation

The number of degrees a figure rotates

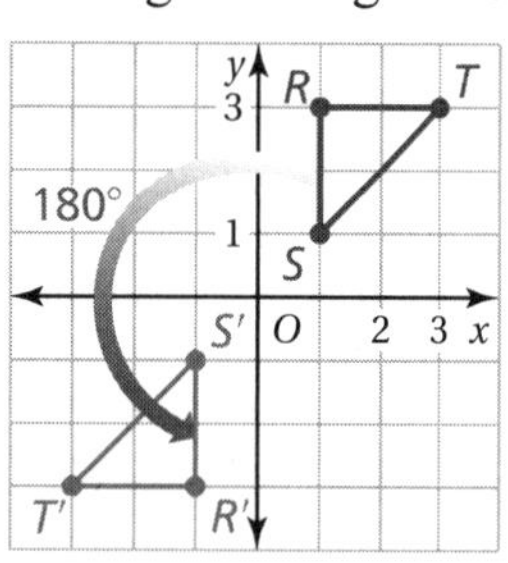

ΔRST has been rotated 180° to $\Delta R'S'T'$.

area

The amount of surface covered by a figure; Area is measured in square units such as square feet (ft^2) or square meters (m^2).

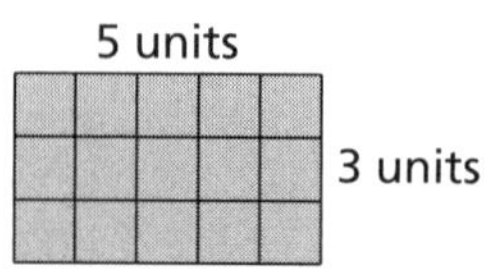

$$A = 5 \times 3 = 15 \text{ square units}$$

arithmetic sequence

A sequence in which the difference between consecutive terms is the same; This difference is called the common difference.

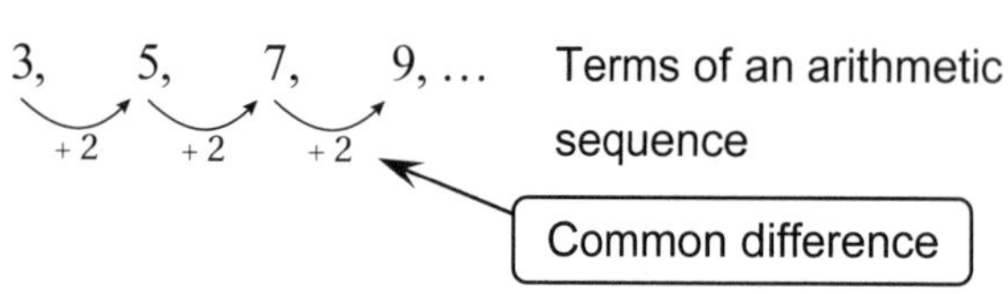

Associative Properties of Addition and Multiplication

Changing the grouping of addends or factors does not change the sum or product.

$$(3 + 4) + 5 = 3 + (4 + 5)$$

$$(3 \bullet 4) \bullet 5 = 3 \bullet (4 \bullet 5)$$

bar graph

A graph in which the lengths of bars are used to represent and compare data

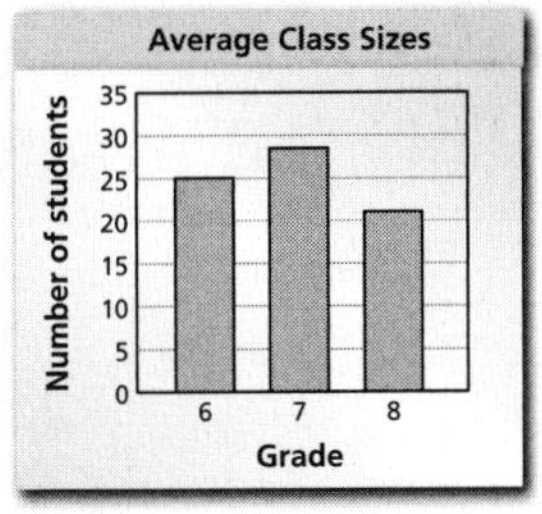

base (of a power)

The base of a power is the common factor.

See power.

capacity

The amount a container can hold

center of dilation

A fixed point with respect to which a figure is dilated

See dilation.

center of rotation

A fixed point about which a figure is rotated

See rotation.

circle

The set of all points in a plane that are the same distance from a point called the center

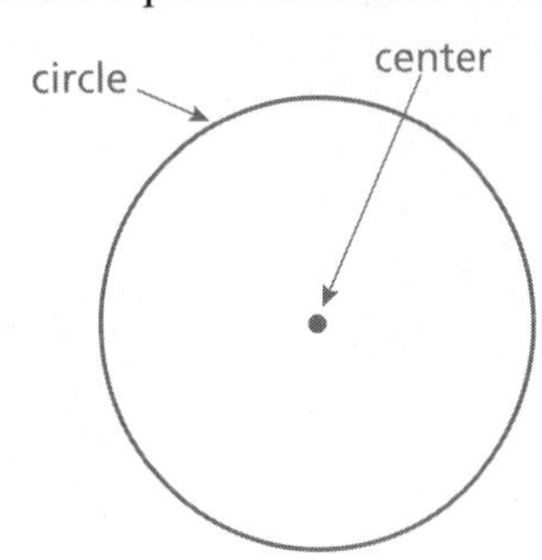

circle graph

Displays data as sections of a circle; The circle represents all of the data. Each section represents part of the data. The sum of the angle measures in a circle graph is $360°$.

Favorite Fruit

circumference

The distance around a circle

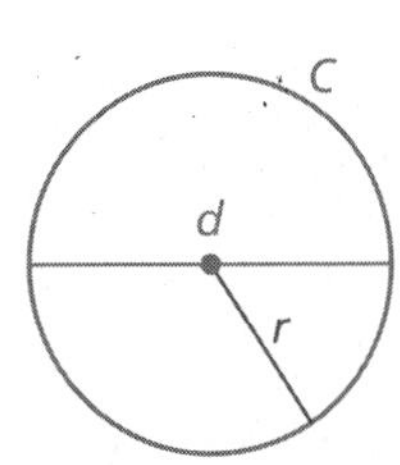

coefficient

The numerical factor of a term that contains a variable

In the algebraic expression $6k + 8$, 6 is the coefficient of the term $6k$.

common difference

The difference between consecutive terms of an arithmetic sequence

See arithmetic sequence.

common factor

A factor that is shared by two or more whole numbers

2 is a common factor of 8 and 10.

common ratio

The ratio between consecutive terms of a geometric sequence

See geometric sequence.

Commutative Properties of Addition and Multiplication

Changing the order of addends or factors does not change the sum or product.

$$2 + 8 = 8 + 2$$
$$2 \bullet 8 = 8 \bullet 2$$

composite number

A whole number greater than 1 that has factors other than itself and 1

12 is a composite number because the factors of 12 are 1, 2, 3, 4, 6, and 12.

compound event

An event that consists of two or more events

Spinning a spinner and flipping a coin

congruent

Having the same size and shape

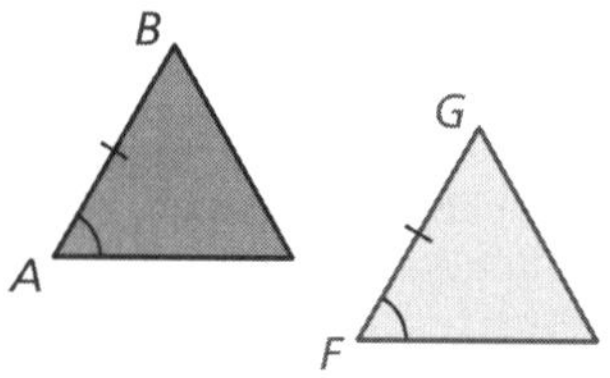

$\angle A$ is congruent to $\angle F$.

Side AB is congruent to Side FG.

congruent (figures)

Figures that have the same size and shape

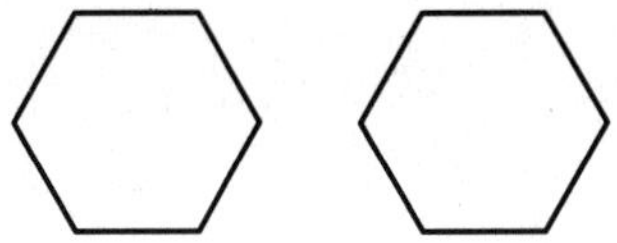

consecutive terms (of a sequence)

Terms that are next to each other in a sequence

Sequence: 1, 3, 5, 7, 9, 11, …

Consecutive terms: 3 and 5, 9 and 11

constant term

A term without a variable

In the expression $2x + 8$, the term 8 is a constant term.

coordinate plane

A coordinate plane is formed by the intersection of a horizontal number line, usually called the x-axis, and a vertical number line, usually called the y-axis.

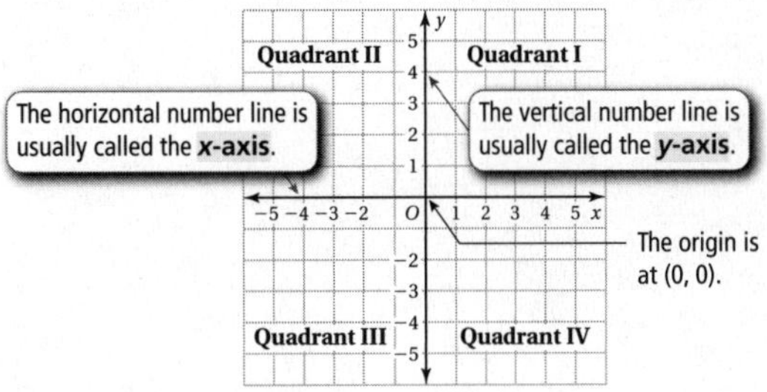

corresponding angles

Matching angles of two similar figures

$\triangle ABC \sim \triangle DEF$

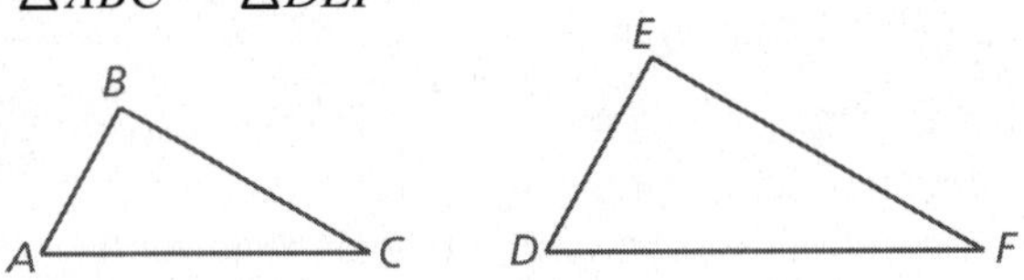

Corresponding angles: $\angle A$ and $\angle D$

$\angle B$ and $\angle E$

$\angle C$ and $\angle F$

corresponding sides

Matching sides of two similar figures

$\triangle ABC \sim \triangle DEF$

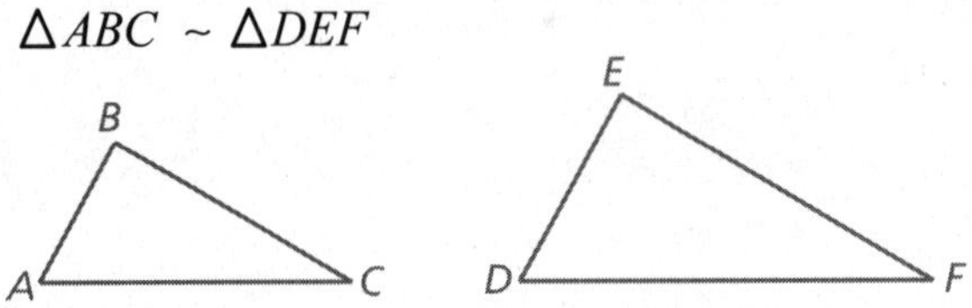

Corresponding sides: Side AB and Side DE

Side BC and Side EF

Side AC and Side DF

cross products

In the proportion $\frac{a}{b} = \frac{c}{d}$, the products $a \bullet d$ and $b \bullet c$ are called cross products.

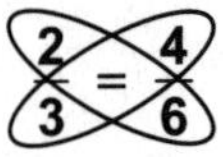

$2 \bullet 6$ and $3 \bullet 4$

Cross Products Property

The cross products of a proportion are equal.

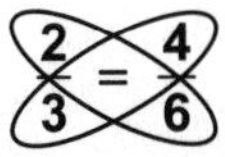

$2 \bullet 6 = 3 \bullet 4$

cube

A rectangular prism with 6 congruent square faces

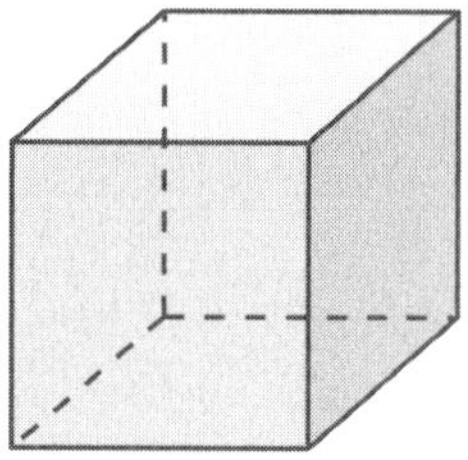

cube(d)

A number cubed is the number raised to an exponent of 3.

2 cubed means 2^3, or 8.

cubic units

The units volume is measured in

cubic feet $\left(\text{ft}^3\right)$, cubic meters $\left(\text{m}^3\right)$

cylinder

A solid that has two parallel, congruent circular bases

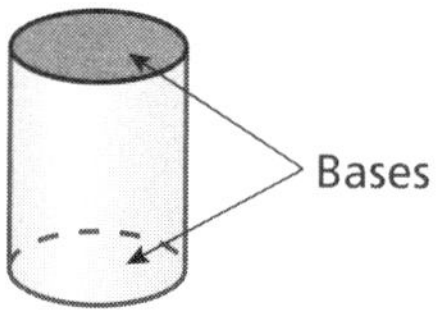

data

Information, often given in the form of numbers or facts

decimal

A number that is written using the base-ten place value system; Each place value is ten times the place value to the right.

The decimal 2.15 represents 2 ones plus 1 tenth plus 5 hundredths, or two and fifteen hundredths.

degree

A unit used to measure angles

90°, 45°, 32°

denominator

The number below the fraction bar in a fraction

In the fraction $\frac{2}{5}$, the denominator is 5.

dependent events

Two events such that the occurrence of one event affects the likelihood that the other event will occur

A bag contains 3 red marbles and 4 blue marbles. You randomly draw a marble, do not replace it, then randomly draw another marble. The events "first marble is blue" and "second marble is red" are dependent events.

diameter (of a circle)

The distance across a circle through the center

See circumference.

difference

The result when one number is subtracted from another number

The difference of 4 and 3 is $4 - 3$, or 1.

dilation

A transformation in which a figure is made larger or smaller with respect to a fixed point called the center of dilation

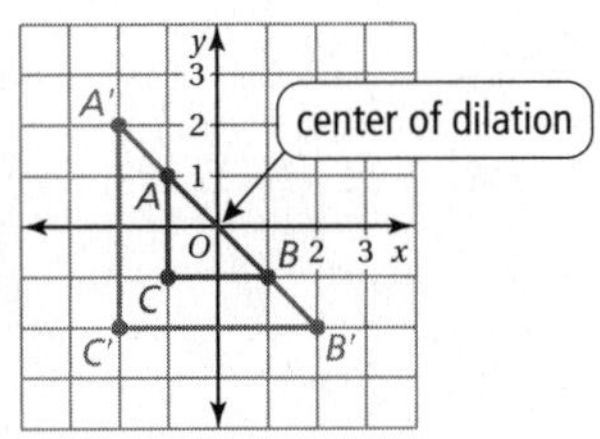

$A'B'C'$ is a dilation of ABC with respect to the origin. The scale factor is 2.

Distributive Property

To multiply a sum or difference by a number, multiply each number in the sum or difference by the number outside the parentheses. Then evaluate.

$$3(2 + 9) = 3(2) + 3(9)$$
$$3(2 - 9) = 3(2) - 3(9)$$

Division Property of Equality

Dividing each side of an equation by the same number produces an equivalent equation.

$$-3y = 18$$
$$\frac{-3y}{-3} = \frac{18}{-3}$$
$$y = -6$$

Division Property of Inequality

If you divide each side of an inequality by the same positive number, the inequality remains true.

If you divide each side of an inequality by the same negative number, the direction of the inequality symbol must be reversed for the inequality to remain true.

$$4x > -12 \qquad -5x > 30$$
$$\frac{4x}{4} > \frac{-12}{4} \qquad \frac{-5x}{-5} < \frac{30}{-5}$$
$$x > -3 \qquad x < -6$$

enlargement

A dilation with a scale factor greater than 1

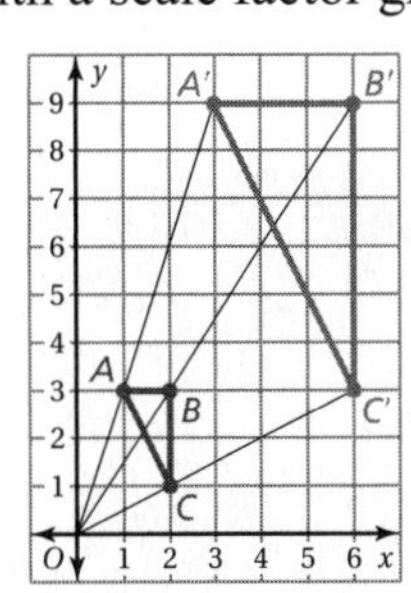

$A'B'C'$ is an enlargement of ABC.

equation

A mathematical sentence that uses an equal sign to show that two expressions are equal

$4x = 16, a + 7 = 21$

equivalent equations

Equations that have the same solution(s)

$2x - 8 = 0$ and $2x = 8$

equivalent expressions

Expressions with the same value

$7 + 4, 4 + 7$

equivalent fractions

Fractions that represent the same number

$\frac{2}{4}$ and $\frac{9}{18}$ are equivalent fractions that both represent $\frac{1}{2}$.

estimate

noun: An approximate solution to a problem

2π is about 6.28.

verb: To find an approximate solution to a problem

You can estimate the sum of $98 + 53$ as $100 + 50$, or 150.

evaluate (an algebraic expression)

Substitute a number for each variable in an algebraic expression. Then use the order of operations to find the value of the numerical expression.

Evaluate $3x + 5$ when $x = 6$.

$$\begin{aligned} 3x + 5 &= 3(6) + 5 \\ &= 18 + 5 \\ &= 23 \end{aligned}$$

event

A collection of one or more outcomes of an action

Flipping heads on a coin

experiment

An activity with varying results

Rolling a number cube

experimental probability

Probability that is based on repeated trials of an experiment

$$P(\text{event}) = \frac{\text{number of times the event occurs}}{\text{total number of trials}}$$

A basketball player makes 19 baskets in 28 attempts. The experimental probability that the player makes a basket is $\frac{19}{28} = 68\%$.

exponent

The exponent of a power indicates the number of times a base is used as a factor.

See power.

expression

A mathematical phrase containing numbers, operations, and/or variables

$12 + 6, 18 + 3 \times 4$

$8 + x, 6 \times a - b$

faces of a solid

The polygons that form a solid figure

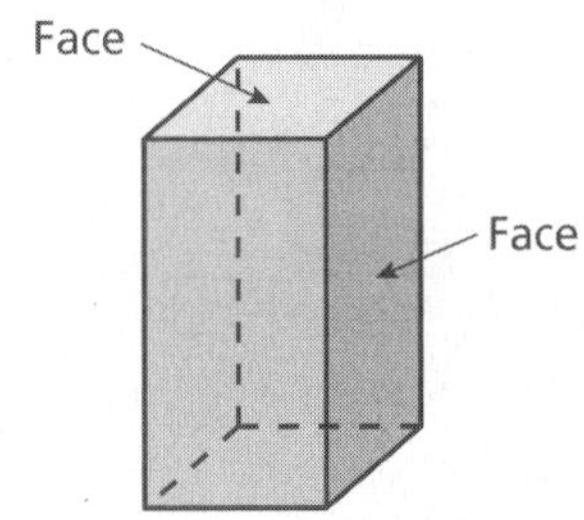

factor

When whole numbers other than zero are multiplied together, each number is a factor of the product.

$2 \times 3 \times 4 = 24$, so 2, 3, and 4 are factors of 24.

fair experiment

An experiment in which all of the possible outcomes are equally likely

The spinner is equally likely to land on 1 or 2. The spinner is fair.

The spinner is more likely to land on 1 than on either 2 or 3. The spinner is *not* fair.

favorable outcome

The outcomes of a specific event

When rolling a number cube, the favorable outcomes for the event "rolling an even number" are 2, 4, and 6.

formula

An equation that tells you how one variable is related to one or more other variables

$A = \ell w$ is the formula for the area of a rectangle.

fraction

A number in the form $\frac{a}{b}$, where $b \neq 0$

$$\frac{1}{2}, \frac{5}{9}$$

frequency table

A table used to count how many times data values occur in intervals

Pairs of shoes	Frequency
1–5	11
6–10	4
11–15	0
16–20	3
21–25	6

frieze

A horizontal band that runs at the top of a building; A frieze is often decorated with a design that repeats.

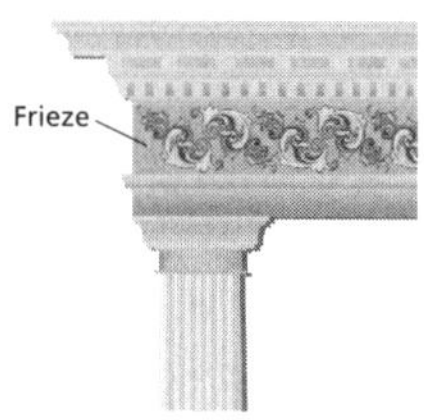

function

A relationship that pairs each input with exactly one output

The ordered pairs $(0, 1)$, $(1, 2)$, $(2, 4)$, and $(3, 6)$ represent a function.

Ordered Pairs

$(0, 1)$
$(1, 2)$
$(2, 4)$
$(3, 6)$

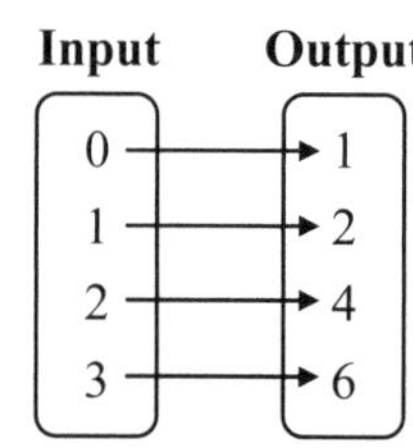

function rule

An equation that describes the relationship between inputs and outputs

The function rule "the output is three less than the input" is represented by the equation $y = x - 3$.

Fundamental Counting Principle

An event M has m possible outcomes and event N has n possible outcomes. The total number of outcomes of event M followed by event N is $m \times n$.

You have 7 shirts, 5 pairs of pants, and 2 pairs of shoes. You can make $7 \times 5 \times 2 = 70$ different outfits.

geometric sequence

A sequence in which the ratio between consecutive terms is the same; This ratio is called the common ratio.

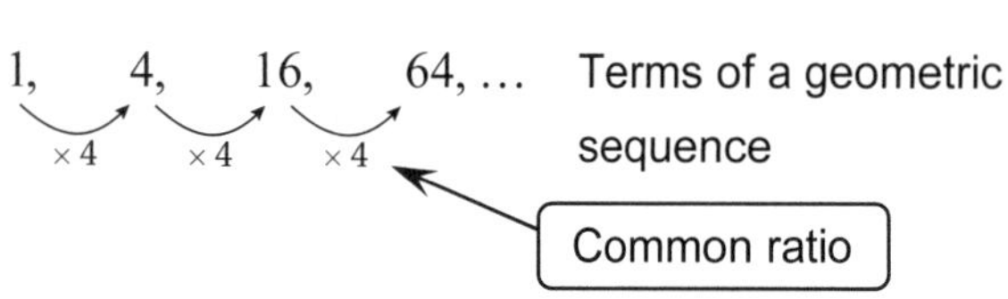

graph (of a function)

A representation of all the points that are solutions of a function rule

The graph of $y = x + 2$ is shown.

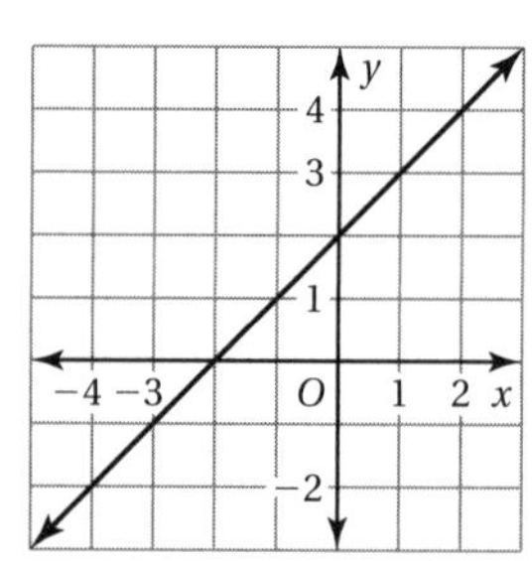

greatest common factor (GCF)

The largest of the common factors of two or more nonzero whole numbers

The common factors of 12 and 20 are 1, 2, and 4. So the GCF of 12 and 20 is 4.

histogram

A bar graph that shows the frequency of data values in intervals of the same size; The height of a bar represents the frequency of the values in the interval. There are no spaces between bars.

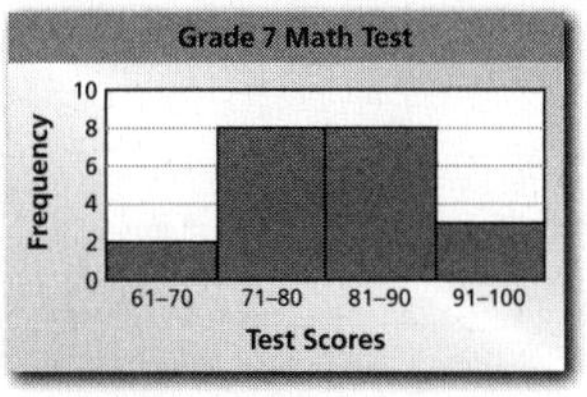

image

The new figure formed by a transformation

See translation, reflection, rotation, and dilation.

improper fraction

A fraction in which the numerator is greater than or equal to the denominator

$$\frac{5}{4}, \frac{9}{9}$$

independent events

Two events such that the occurrence of one event does not affect the likelihood that the other event will occur

You flip a coin and roll a number cube. The events "flipping tails" and "rolling a 4" are independent events.

indirect measurement

Using similar figures to find a missing measure that is difficult to find directly

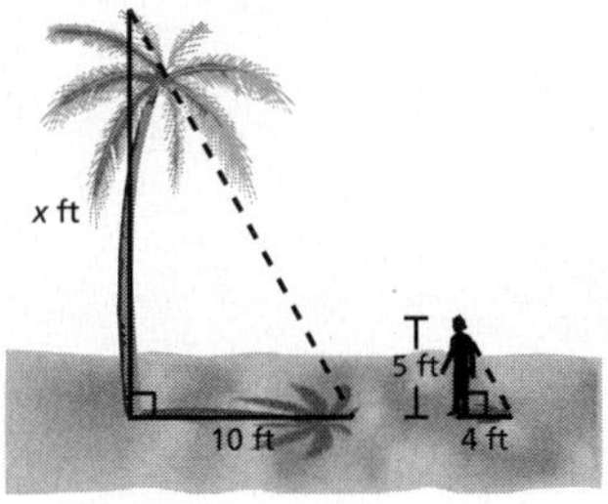

$$\frac{x}{10} = \frac{5}{4}$$

$$10 \bullet \frac{x}{10} = 10 \bullet \frac{5}{4}$$

$$x = 12.5$$

The tree is 12.5 feet tall.

inequality

A mathematical sentence that compares expressions; It contains the symbols $<$, $>$, $\leq$, or $\geq$.

$$x - 4 < 14,\ x + 5 \geq -12$$

input

A number on which a function operates

See function.

input-output table

A table that lists the output of a function for each input

Input, x	Output, y
1	3
2	4
3	5
4	6

integers

The set of whole numbers and their opposites

$$\ldots -3, -2, -1, 0, 1, 2, 3, \ldots$$

inverse operations

Operations that "undo" each other, such as addition and subtraction or multiplication and division

isosceles triangle

A triangle that has at least two congruent sides

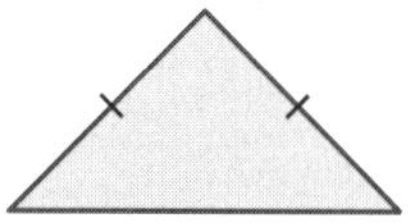

kite

A quadrilateral with two pairs of congruent adjacent sides and opposite sides that are not congruent

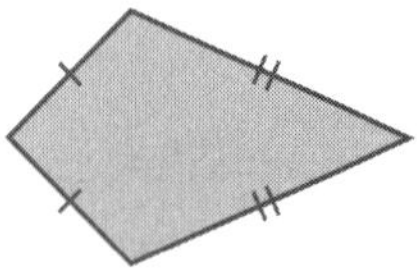

lateral face

Any face or surface that is not a base

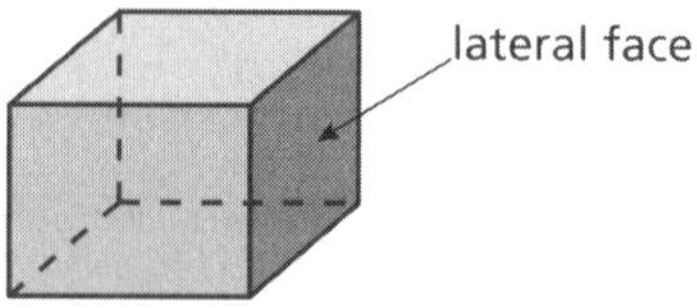

lateral surface area

The surface area of a figure excluding the area of its base(s)

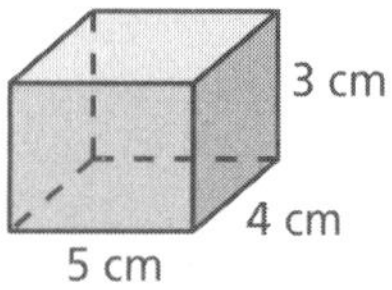

$$\text{Lateral surface area} = 2(4)(3) + 2(5)(3)$$
$$= 24 + 30 = 54 \text{ cm}^2$$

leaf

Digit or digits on the right of a stem-and-leaf plot

See stem-and-leaf plot.

least common denominator (LCD)

The least common multiple of the denominators of two or more fractions

The least common denominator of $\frac{3}{4}$ and $\frac{5}{6}$ is the least common multiple of 4 and 6, or 12.

least common multiple (LCM)

The smallest of the common multiples of two or more nonzero whole numbers

Multiples of 10: 10, 20, 30, 40, …
Multiples of 15: 15, 30, 45, 60, …

The least common multiple of 10 and 15 is 30.

like terms

Terms of an algebraic expression that have the same variables raised to the same exponents

4 and 8, $2x$ and $7x$

line

A set of points that extends without end in two opposite directions

line graph

A type of graph in which points representing data pairs are connected by line segments

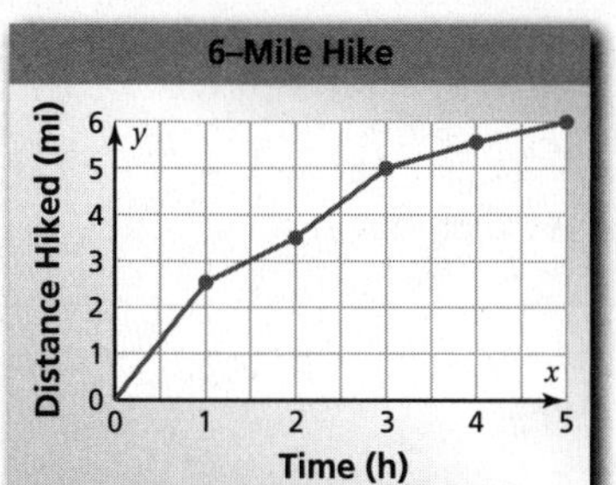

line of reflection

A line that a figure is reflected in to create a mirror image of the original figure

See reflection.

line segment

Part of a line that consists of two points, called endpoints, and all of the points on the line between the endpoints

linear function

A function whose graph is a straight line

$y = x - 1$

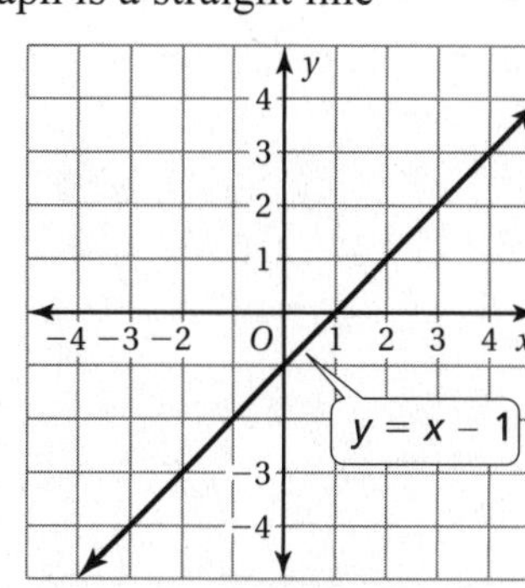

mapping diagram

A way to represent a function

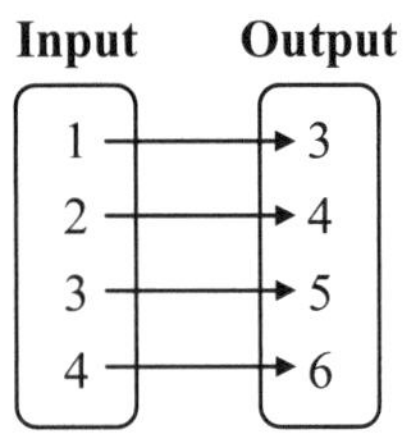

mean

The sum of the values in a data set divided by the number of data values

The mean of the values 7, 4, 8, and 9 is

$$\frac{7 + 4 + 8 + 9}{4} = \frac{28}{4} = 7.$$

median

For a data set with an odd number of ordered values, the median is the middle data value. For a data set with an even number of ordered values, the median is the mean of the two middle values.

The median of the data set 24, 25, 29, 33, 38 is 29 because 29 is the middle value.

metric system

Decimal system of measurement, based on powers of 10, that contains units for length, capacity, and mass

centimeter, meter, liter, kilogram

mixed number

A number that has a whole number part and a fraction part

$$3\frac{1}{2}, 6\frac{2}{3}$$

mode

The data value or values that occur most often; Data can have one mode, more than one mode, or no mode.

The modes of the data set 3, 4, 4, 7, 7, 9, 12 are 4 and 7 because they occur most often.

Multiplication Properties of Zero and One

The product of any number and 0 is 0.

The product of any number and 1 is that number.

$$5 \bullet 0 = 0$$
$$6 \bullet 1 = 6$$

Multiplication Property of Equality

Multiplying each side of an equation by the same number produces an equivalent equation.

$$\frac{x}{3} = -6$$
$$3 \bullet \frac{x}{3} = 3 \bullet (-6)$$
$$x = -18$$

Multiplication Property of Inequality

If you multiply each side of an inequality by the same positive number, the inequality remains true.

If you multiply each side of an inequality by the same negative number, the direction of the inequality symbol must be reversed for the inequality to remain true.

$$\frac{x}{2} < -9 \qquad\qquad \frac{x}{-6} < 3$$

$$2 \bullet \frac{x}{2} < 2 \bullet (-9) \qquad\qquad -6 \bullet \frac{x}{-6} > -6 \bullet 3$$

$$x < -18 \qquad\qquad x > -18$$

negative number

A number less than 0

$-0.25, -10, -500$

net

A two-dimensional representation of a solid

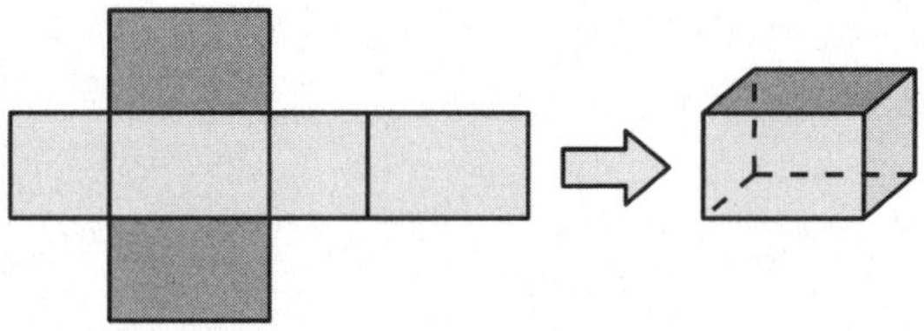

number line

A line whose points are associated with numbers that increase from left to right

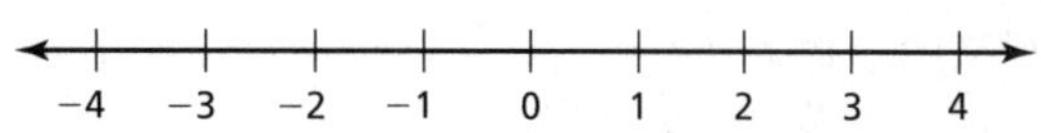

numerator

The number above the fraction bar in a fraction

In the fraction $\frac{2}{5}$, the numerator is 2.

numerical expression

An expression that contains only numbers and operations

$12 + 6, 18 + 3 \times 4$

obtuse angle

An angle whose measure is greater than $90°$ and less than $180°$

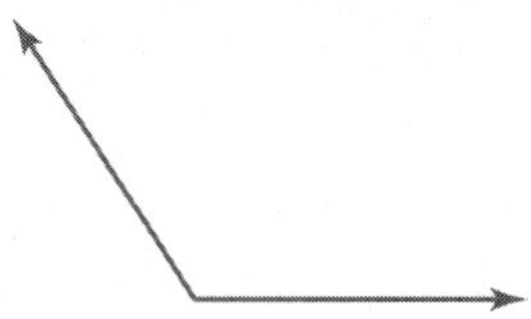

opposites

Two numbers that are the same distance from 0, but on opposite sides of 0

-3 and 3 are opposites.

order of operations

The order in which to perform operations when evaluating expressions with more than one operation

To evaluate $5 + 2 \times 3$, you perform the multiplication before the addition.

ordered pair

A pair of numbers (x, y) used to locate a point in a coordinate plane; The first number is the x-coordinate, and the second number is the y-coordinate.

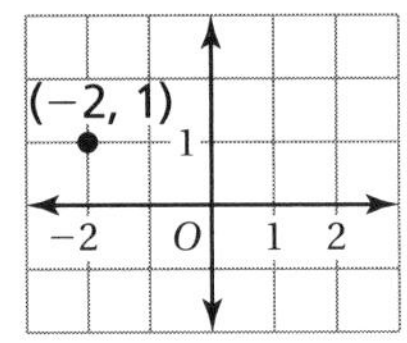

The x-coordinate of the point $(-2, 1)$ is -2, and the y-coordinate is 1.

origin

The point, represented by the ordered pair (0, 0), where the x-axis and the y-axis meet in a coordinate plane

See coordinate plane.

outcomes

The possible results of an action

The outcomes of flipping a coin are heads and tails.

outlier

A data value that is much greater or much less than the other values

In the data set 23, 42, 33, 117, 36, and 40, the outlier is 117.

output

A number produced by evaluating a function using a given input

See function.

parallel (lines)

Two lines in the same plane that do not intersect

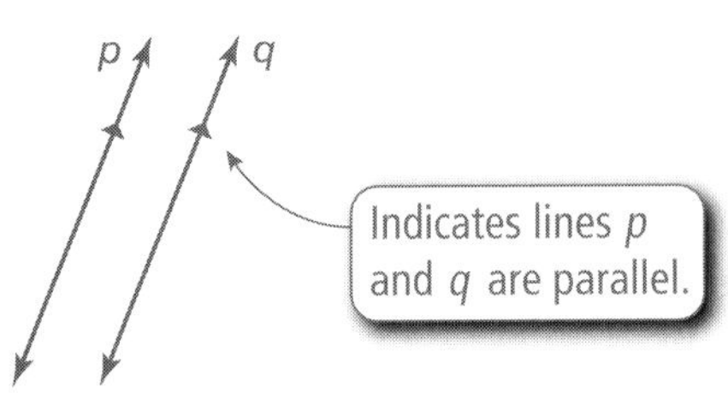

parallelogram

A quadrilateral with two pairs of parallel sides

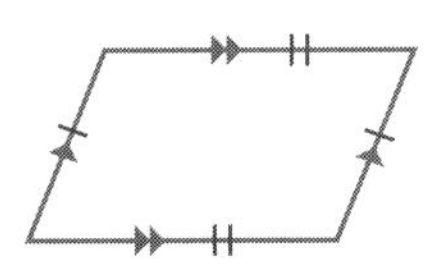

percent

The number of parts per one hundred

$$37\% = 37 \text{ out of } 100 = \frac{37}{100}$$

perfect square

A number with integers as its square roots

16, 25, 81

perimeter

The distance around a polygon; Perimeter is measured in linear units such as feet (ft) or meters (m).

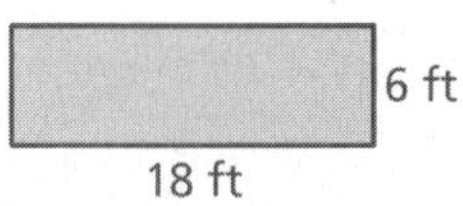

$$\text{Perimeter} = 18 + 6 + 18 + 6 = 48 \text{ ft}$$

pi (π)

The ratio of the circumference of a circle to its diameter

You can use 3.14 or $\frac{22}{7}$ to approximate π.

place value

The place value of each digit in a number depends on its position within the number.

In 521, 5 is in the hundreds place and has a value of 500.

point

A position in space represented with a dot

polygon

A closed plane figure made up of three or more line segments that intersect only at their endpoints

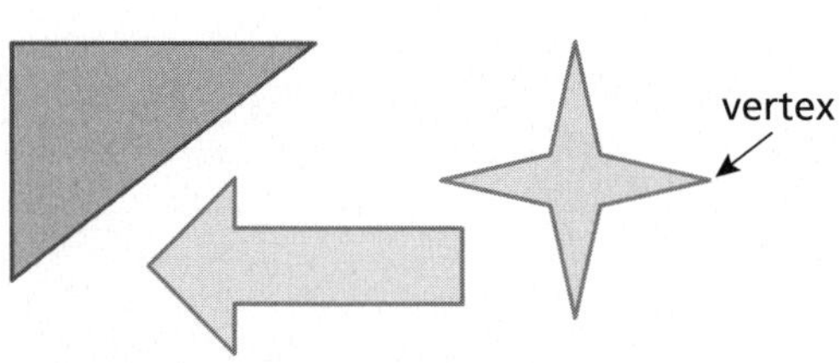

polyhedron

A three-dimensional figure whose faces are all polygons

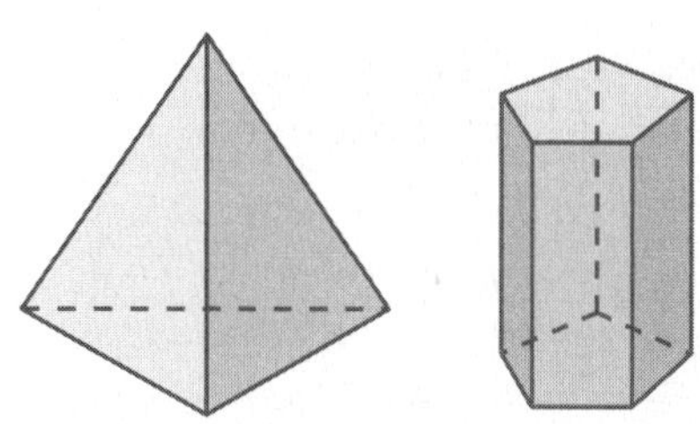

positive number

A number greater than 0

0.5, 2, 100

power

A product of repeated factors

base → $\left(\frac{1}{2}\right)^5$ ← exponent

$$\left(\frac{1}{2}\right)^5 = \frac{1}{2} \bullet \frac{1}{2} \bullet \frac{1}{2} \bullet \frac{1}{2} \bullet \frac{1}{2}$$

Power $\quad \frac{1}{2}$ is used as a factor 5 times.

prism

A polyhedron that has two parallel, congruent bases; The other faces are parallelograms.

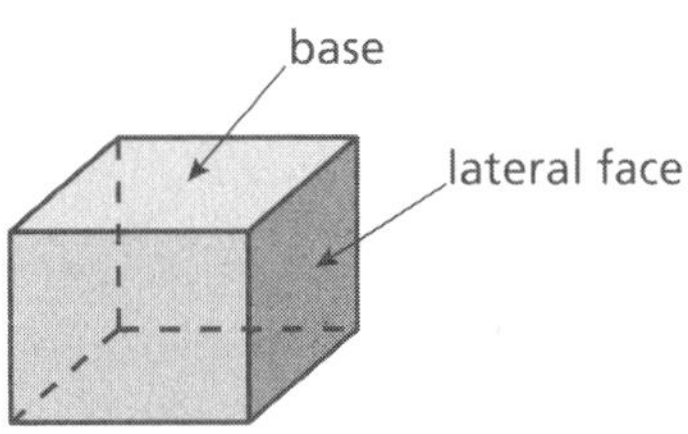

probability

A number from 0 to 1 that measures the likelihood that an event will occur

See experimental probability and theoretical probability.

product

The result when two or more numbers are multiplied

The product of 4 and 3 is 4×3, or 12.

proportion

An equation stating that two ratios are equivalent

$$\frac{3}{4} = \frac{12}{16}$$

proportional

Two quantities that form a proportion are proportional.

Because $\frac{3}{4}$ and $\frac{12}{16}$ form a proportion, $\frac{3}{4}$ and $\frac{12}{16}$ are proportional.

protractor

A tool used to measure angles

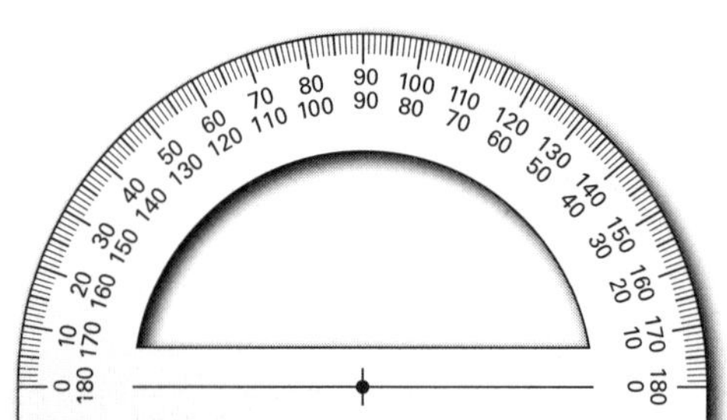

quadrants

The four regions created by the intersection of the x-axis and the y-axis in a coordinate plane

See coordinate plane.

quadrilateral

A polygon with four sides

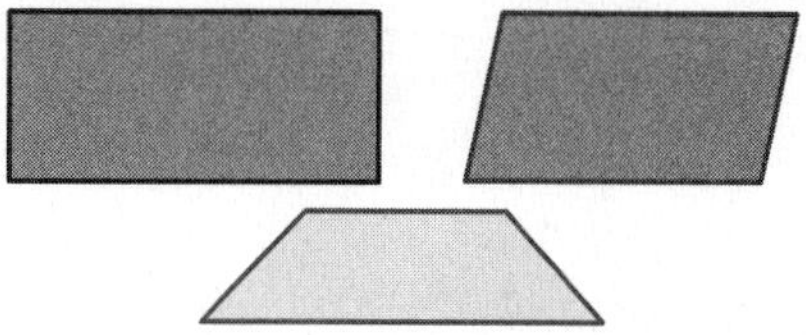

quotient

The result of a division

The quotient of 10 and 5 is $10 \div 5$, or 2.

radical sign

The symbol $\sqrt{\ }$ which is used to represent a square root

$$\sqrt{25} = 5$$

$$-\sqrt{49} = -7$$

$$\pm\sqrt{100} = \pm 10$$

radicand

The number under a radical sign

The radicand of $\sqrt{25}$ is 25.

radius (of a circle)

The distance from the center of a circle to any point on the circle

See circumference.

range (of a data set)

The difference between the greatest value and the least value of a data set; The range describes how spread out the data are.

The range of the data set 12, 16, 18, 22, 27, 35 is $35 - 12 = 23$.

rate

A ratio of two quantities with different units

You read 3 books every 2 weeks.

ratio

A comparison of two quantities using division; The ratio of a to b $(\text{where } b \neq 0)$ can be written as a to b, $a : b$, or $\frac{a}{b}$.

4 to 1, 4 : 1, or $\frac{4}{1}$

rational number

A number that can be written as $\frac{a}{b}$ where a and b are integers and $b \neq 0$

$3 = \frac{3}{1}$, $\quad -\frac{2}{5} = \frac{-2}{5}$

$0.25 = \frac{1}{4}$, $\quad 1\frac{1}{3} = \frac{4}{3}$

ray

A part of a line that has one endpoint and extends without end in one direction

reciprocals

Two numbers whose product is 1

Because $\frac{4}{5} \times \frac{5}{4} = 1$, $\frac{4}{5}$ and $\frac{5}{4}$ are reciprocals.

rectangle

A parallelogram with four right angles

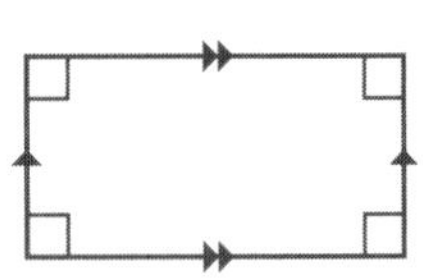

rectangular prism

A three-dimensional figure that has 6 rectangular faces

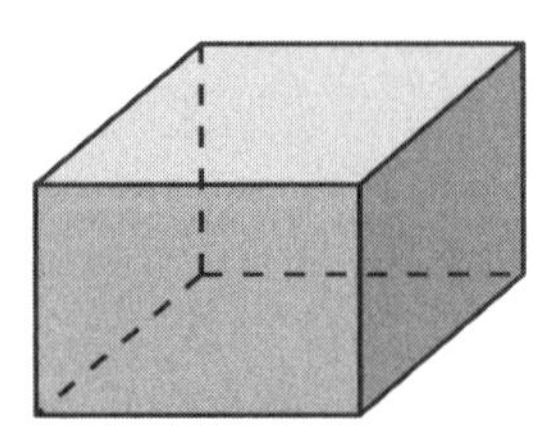

reduction

A dilation with a scale factor greater than 0 and less than 1

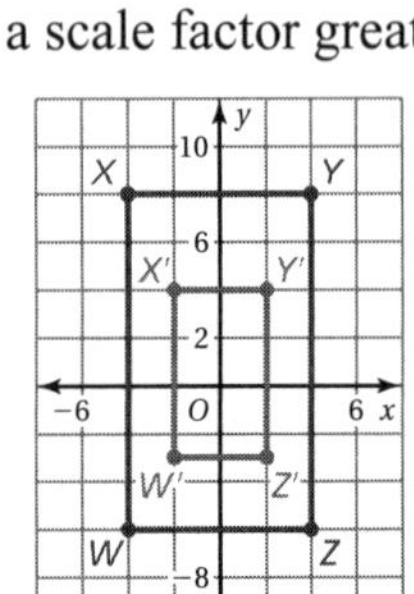

$W'X'Y'Z'$ is a reduction of $WXYZ$.

reflection

A transformation in which a figure is reflected in a line called the line of reflection; A reflection creates a mirror image of the original figure.

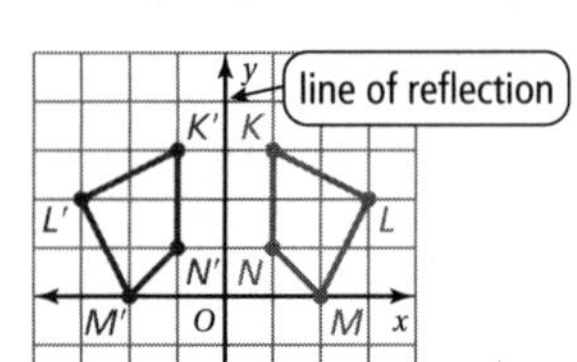

$K'L'M'N'$ is a reflection of $KLMN$ over the y-axis.

remainder

If a divisor does not divide a dividend evenly, the remainder is the whole number left over after the division.

$$\begin{array}{r} 4 \\ 7\overline{)30} \\ \underline{28} \\ 2 \end{array} \quad \text{R}2 \quad \text{The remainder is 2.}$$

repeating decimal

A decimal that has a pattern that repeats

$0.555... = 0.\overline{5}$

$1.727272... = 1.\overline{72}$

rhombus

A parallelogram with four sides of equal length

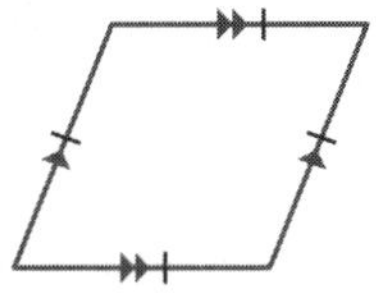

right angle

An angle whose measure is $90°$

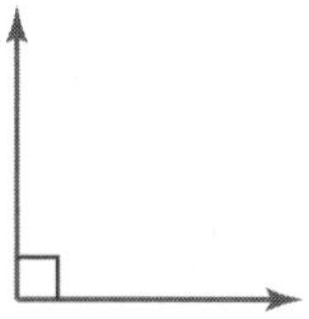

rotation

A transformation in which a figure is rotated about a point called the center of rotation

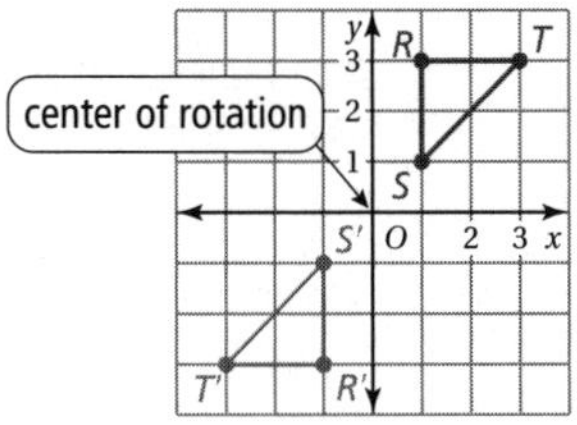

ΔRST has been rotated about the origin O to $\Delta R'S'T'$.

rotational symmetry

A figure has rotational symmetry if a turn of $180°$ or less produces an image that fits exactly on the original figure.

The figure has $60°$ rotational symmetry.

round

To approximate a number to a given place value

132 rounded to the nearest ten is 130.

scale

A ratio that compares the measurements of a drawing or model with the actual measurements

12 cm : 1 cm

2 in. : 15 ft

scale drawing

A proportional two-dimensional drawing of an object

A blueprint or a map

scale factor (of a dilation)

The ratio of the side lengths of the image of a dilation to the corresponding side lengths of the original figure

See dilation.

scale factor (of a scale drawing)

A scale without units

See ratio.

scale model

A proportional three-dimensional model of an object

scientific notation

A number is written in scientific notation when it is represented as the product of a factor and a power of 10. The factor must be greater than or equal to 1 and less than 10.

8.3×10^4

4×10^{-3}

sequence

An ordered list of numbers

4, 7, 10, 13, …

2, 4, 8, 16, …

similar figures

Figures that have the same shape but not necessarily the same size; Two figures are similar if corresponding side lengths are proportional, and corresponding angles are congruent.

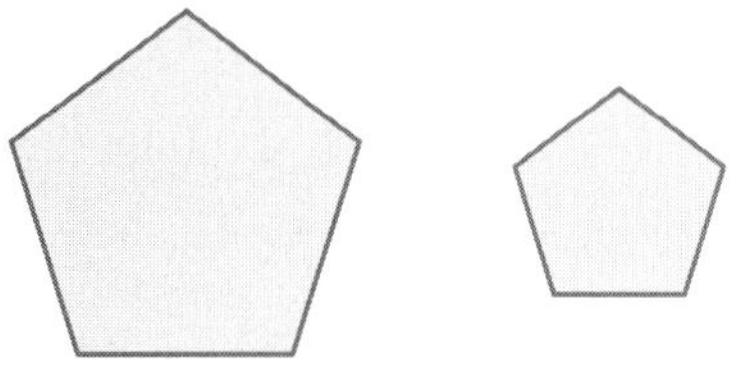

simplest form of an algebraic expression

An algebraic expression is in simplest form if it has no like terms and no parentheses.

$6a + 9a^2, 3t + 5$

simplest form (of a fraction) A fraction is in simplest form if its numerator and denominator have a greatest common factor (GCF) of 1. The simplest form of the fraction $\frac{10}{15}$ is $\frac{2}{3}$.	**slope** A ratio of the change in y (vertical change) to the change in x (horizontal change) between any two points on a line; It is a measure of the steepness of a line. $\text{slope} = \frac{\text{change in } y}{\text{change in } x}$ 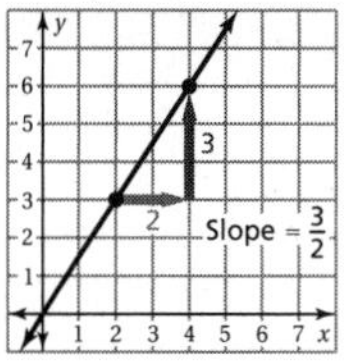
solid A three-dimensional figure *See three-dimensional figure.*	**solution (of an equation)** A value that makes an equation true 6 is the solution of the equation $x - 4 = 2$.
solve a formula Find the value of one variable by substituting numbers for the other variables. $\ell = 6$ in., $w = 10$ in. $P = 2\ell + 2w$ $= 2(6) + 2(10)$ $= 32$ in.	**square** A parallelogram with four congruent sides and four right angles 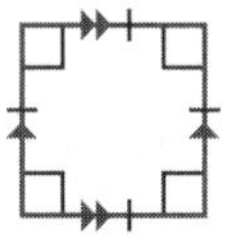
square(d) A number squared is the number raised to an exponent of 2. 5 squared means 5^2, or 25.	**square root** A number that when multiplied by itself, equals the given number The two square roots of 100 are 10 and −10.

stem

Digit or digits on the left of a stem-and-leaf plot

See stem-and-leaf plot.

stem-and-leaf plot

A type of data display that uses the digits of data values to organize a data set; Each data value is broken into a stem (digit or digits on the left) and a leaf (digit or digits on the right).

Test Scores

Stem	Leaf
6	6
7	2 7
8	1 1 3 4 4 6 8 8
9	0 0 0 2 7 8
10	0

Key: 9 | 4 = 94 points

straight angle

An angle whose measure is 180°

Subtraction Property of Equality

Subtracting the same number from each side of an equation produces an equivalent equation.

$$\begin{aligned} w + 5 &= 25 \\ -5 \quad & \quad -5 \\ x &= 20 \end{aligned}$$

Subtraction Property of Inequality

If you subtract the same number from each side of an inequality, the inequality remains true.

$$\begin{aligned} x + 7 &> -20 \\ -7 \quad & \quad -7 \\ x &> -27 \end{aligned}$$

sum

The result when two or more numbers are added

The sum of 4 and 3 is $4 + 3$, or 7.

surface area

The sum of the areas of all the faces or surfaces of a three-dimensional figure

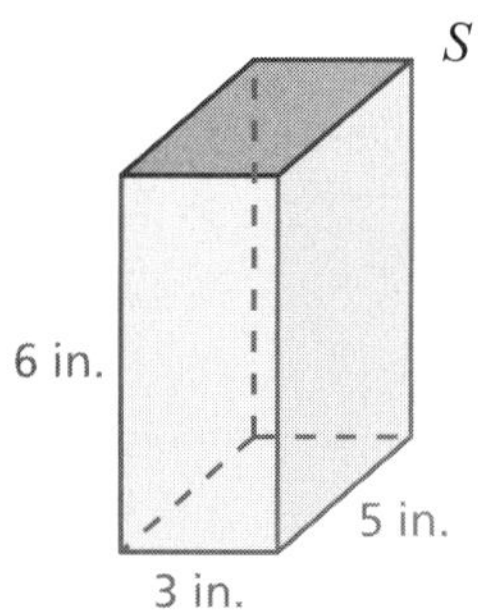

$$\begin{aligned} S &= 2\ell w + 2\ell h + 2wh \\ &= 2(3)(5) + 2(3)(6) + 2(5)(6) \\ &= 30 + 36 + 60 \\ &= 126 \text{ in.}^2 \end{aligned}$$

term (of a sequence)

Each number in a sequence

2, 5, 8, and 11 are terms of the sequence 2, 5, 8, 11, ...

terminating decimal

A decimal that ends

1.5, 2.58, −5.605

terms (of an algebraic expression)

The parts of an expression

The terms of $4x + 7$ are $4x$ and 7.

tessellation

A repeating pattern of congruent plane figures that completely covers a plane with no holes or overlaps

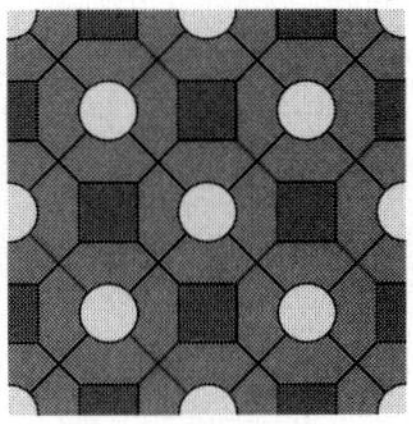

theoretical probability

The ratio of the number of favorable outcomes to the number of possible outcomes when all possible outcomes are equally likely

$$P(\text{event}) = \frac{\text{number of favorable outcomes}}{\text{number of possible outcomes}}$$

When rolling a number cube, the theoretical probability of rolling a 4 is $\frac{1}{6}$.

three-dimensional figure

A figure that has length, width, and depth; also known as a solid

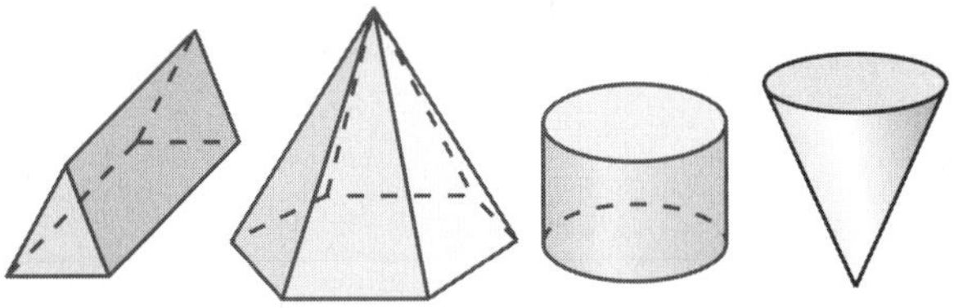

transformation

Changing a figure into another figure

See translation, reflection, rotation, and dilation.

translation

A transformation in which a figure slides but does not turn; Every point of the figure moves the same distance and in the same direction.

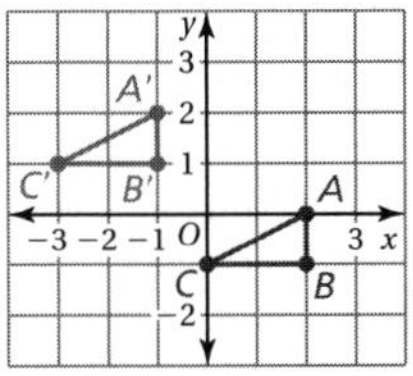

ABC has been translated 3 units left and 2 units up to $A'B'C'$.

trapezoid

A quadrilateral with exactly one pair of parallel sides

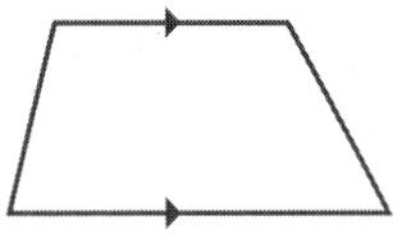

tree diagram

A branching diagram that shows all possible outcomes in a probability experiment

All possible outcomes of tossing a coin three times

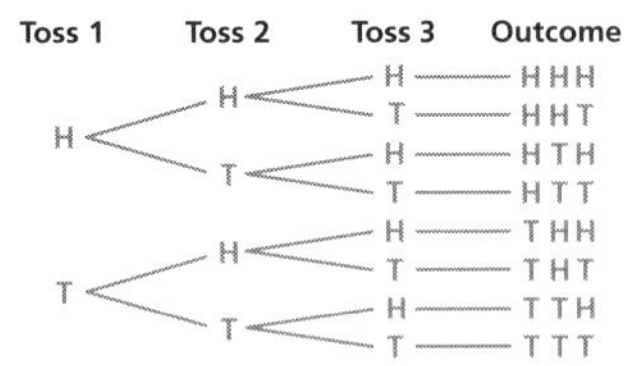

triangle

A polygon with three sides

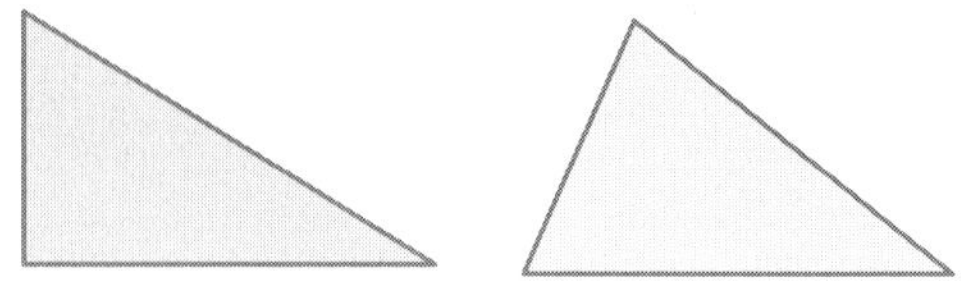

two-dimensional figure

A figure that has only length and width

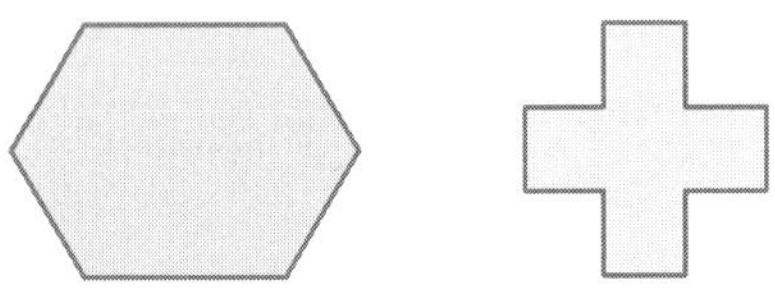

unit rate

A rate with a denominator of 1

The speed limit is 65 miles per hour.

U.S. customary system

System of measurement that contains units for length, capacity, and weight

inches, feet, quarts, gallons, ounces, pounds

variable

A symbol, usually a letter, that represents one or more numbers

x is a variable in $2x + 1$.

variable term

A term that has a variable

In the expression $2x + 8$, the term $2x$ is a variable term.

vertex of a polygon

A point at which two sides of a polygon meet; The plural of vertex is vertices.

See polygon.

volume

A measure of the amount of space that a three-dimensional figure occupies; Volume is measured in cubic units such as cubic feet (ft^3) or cubic meters (m^3).

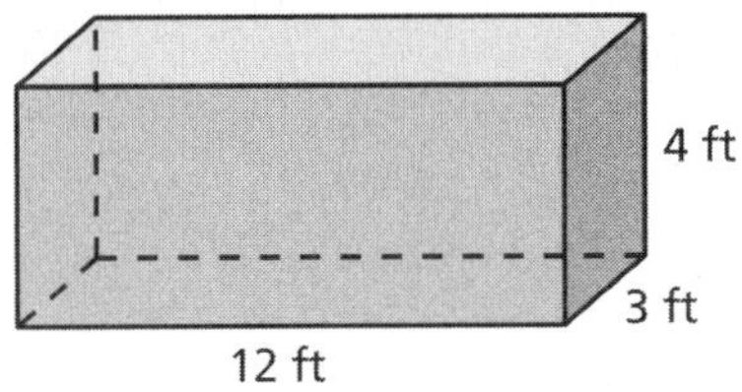

$$\text{Volume} = 12(3) \bullet 4 = 144 \text{ ft}^3$$

whole numbers

The numbers $0, 1, 2, 3, 4, \ldots$

***x*-axis**

The horizontal number line in a coordinate plane

See coordinate plane.

***x*-coordinate**

The first coordinate in an ordered pair, which indicates how many units to move to the left or right

In the ordered pair $(3, 5)$, the x-coordinate is 3.

***y*-axis**

The vertical number line in a coordinate plane

See coordinate plane.

***y*-coordinate**

The second coordinate in an ordered pair, which indicates how many units to move up or down

In the ordered pair $(3, 5)$, the y-coordinate is 5.

Photo Credits

94 ©iStockphoto.com/Michael Flippo; **124** Jean Thompson; **133** *top* ©iStockphoto.com/Viatcheslav Dusaleev; *bottom left* ©iStockphoto.com/Jason Mooy; *bottom center* ©iStockphoto.com/Andres Peiro Palmer; *bottom right* ©iStockphoto.com/Felix Möckel; **188** ©iStockphoto.com/Franck Boston; **192** *Activity 3a* Tom C Amon; *Activity 3b* Tom C Amon, HuHu; *Activity 3c* Olga Gabay; *Activity 3d* NASA/MODIS Rapid Response/Jeff Schmaltz; *Activity 3e* Tom C Amon; **195, 196** ©iStockphoto.com/Kais Tolmats; **200** NASA; **213** *top* John McLaird; *center* Robert Ascento; *bottom* Mark Aplet

Cartoon Illustrations Tyler Stout

Integer Counters*

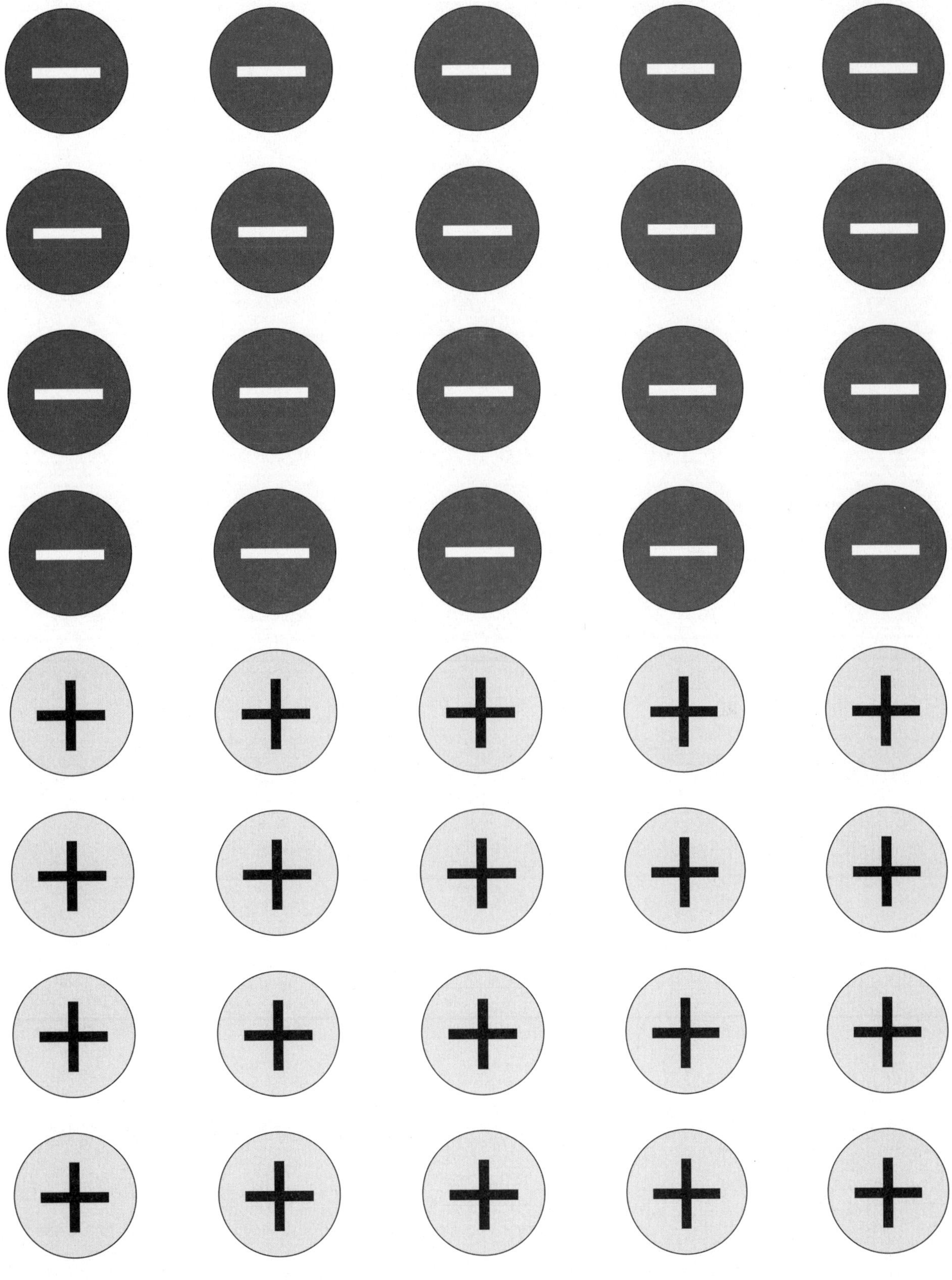

*Available at *BigIdeasMath.com.*

Rongrams – Chapter 2 Section 3*

*Available at *BigIdeasMath.com.*

Algebra Tiles*

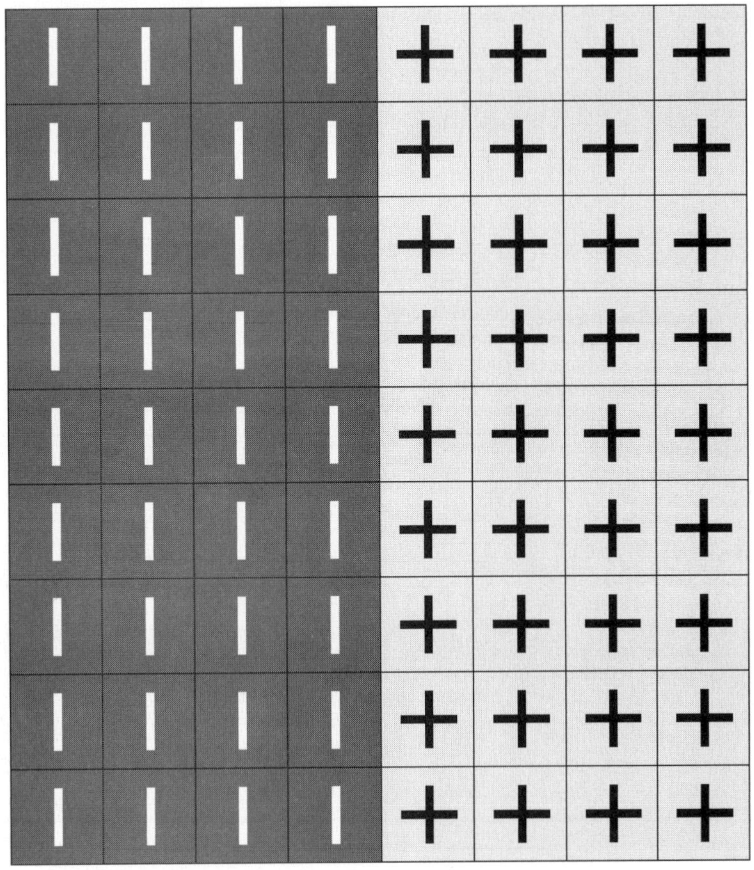

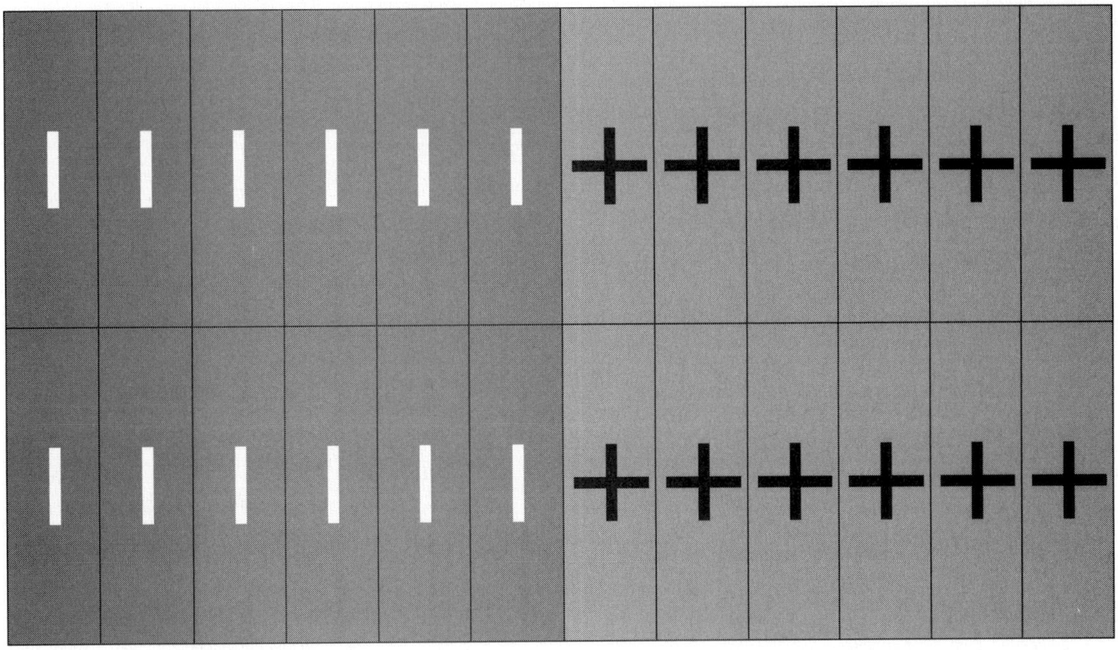

*Available at *BigIdeasMath.com.*

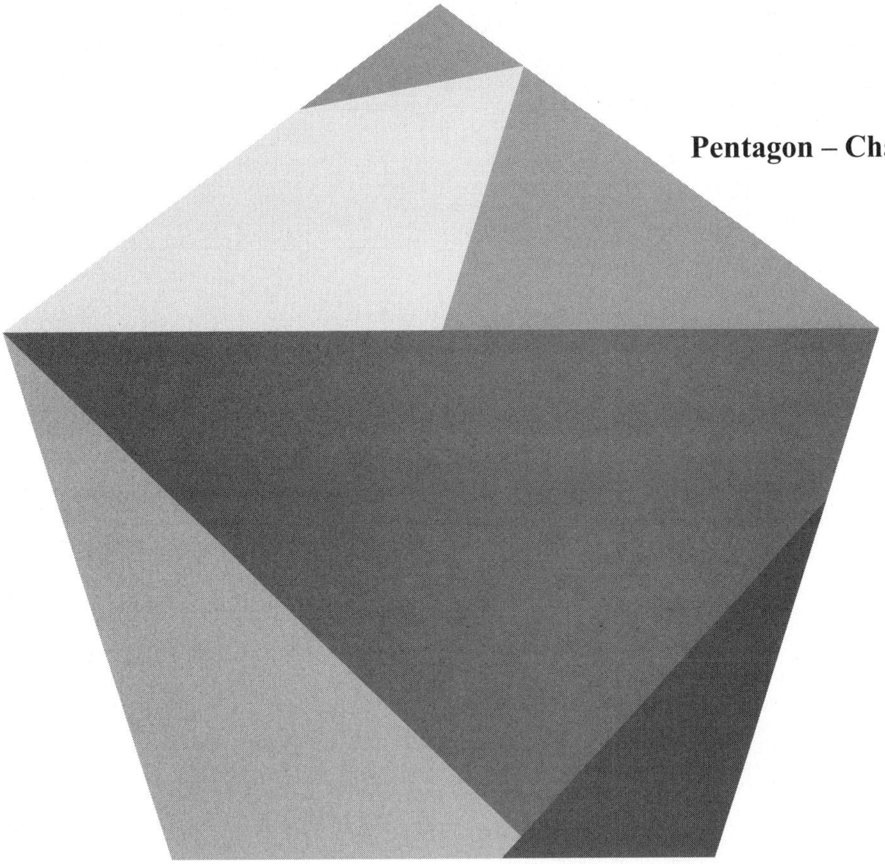

Pentagon – Chapter 3 Section 7*

Hexagon – Chapter 3 Section 7*

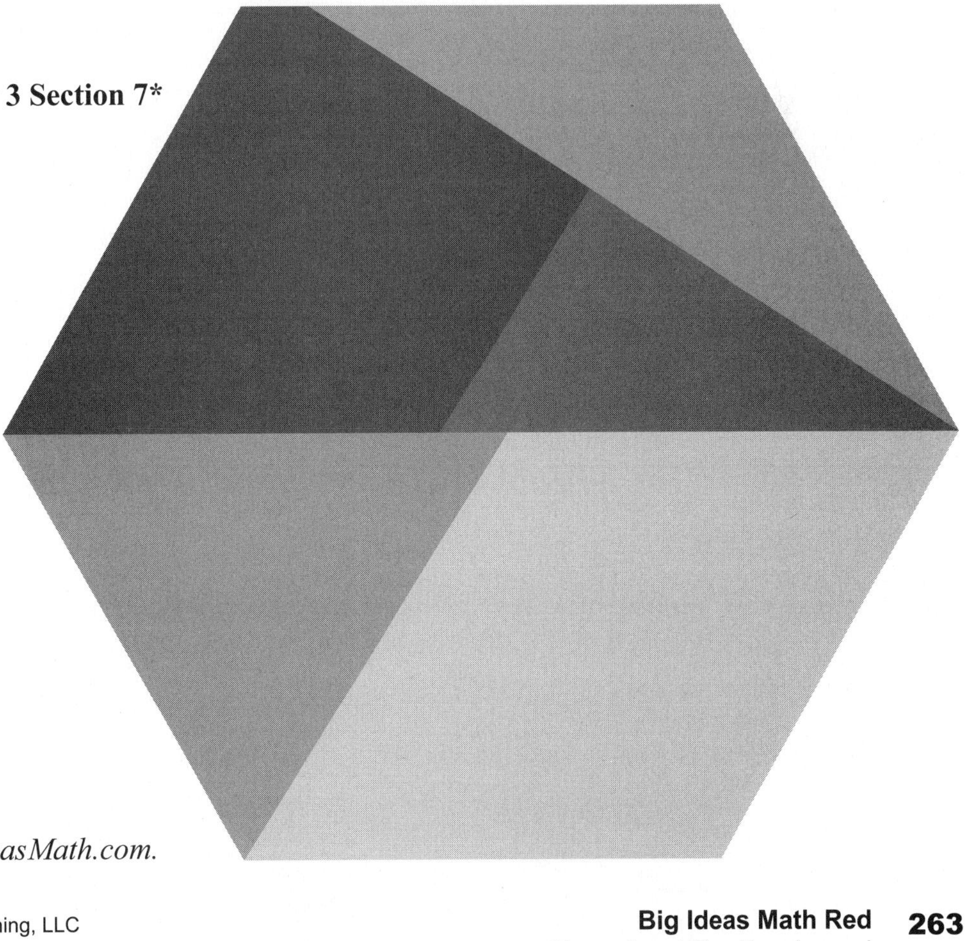

*Available at *BigIdeasMath.com.*

Quadrilaterals – Chapter 6 Section 1*

*Available at *BigIdeasMath.com.*

Pattern Blocks – Chapter 6 Section 3*

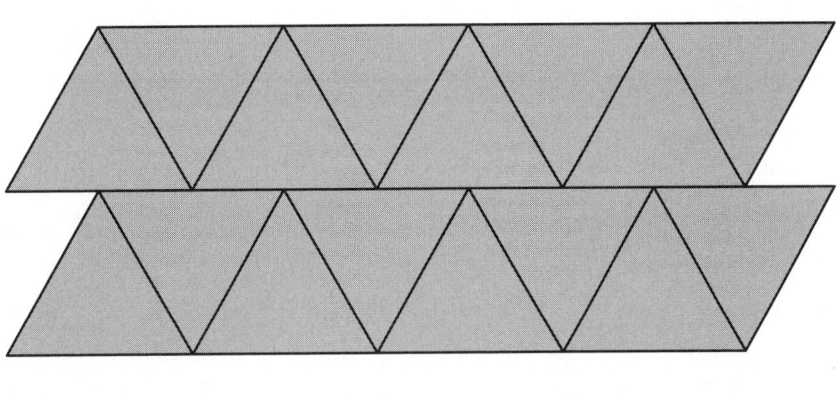

*Available at *BigIdeasMath.com.*

Pattern Blocks – Chapter 6 Section 3*

*Available at *BigIdeasMath.com.*

For use with Chapter 7 Section 2*

a.

b.

c.

d.

*Available at *BigIdeasMath.com.*

For use with Chapter 7 Section 2 (continued)*

e.

f.

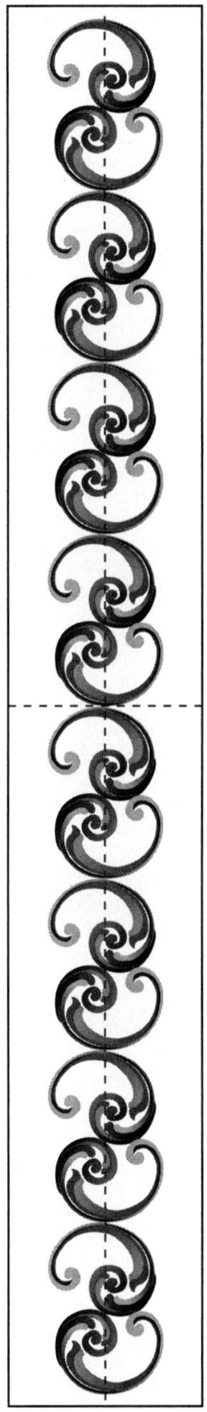

For use with Chapter 7 Section 3*

*Available at *BigIdeasMath.com.*

Spinners – Chapter 10 Section 5*

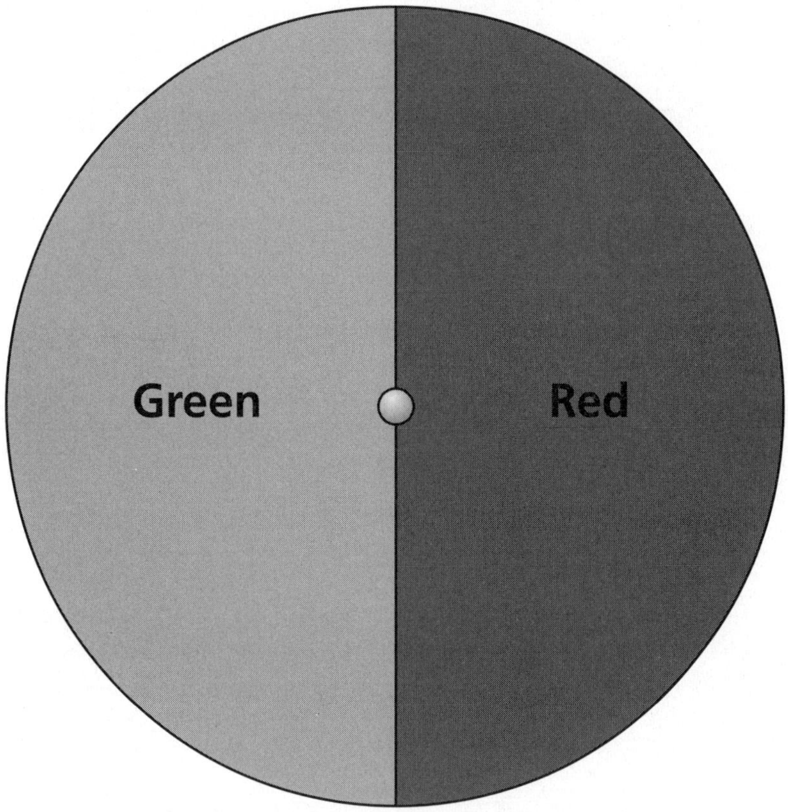

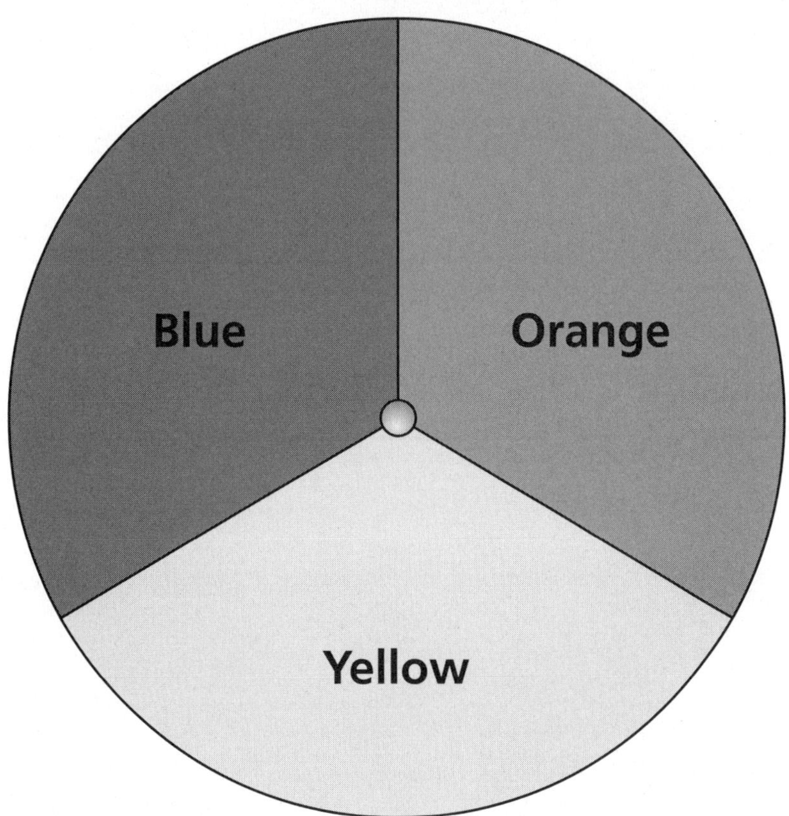

*Available at *BigIdeasMath.com.*